Quality Assurance in LIS Education

Makiko Miwa • Shizuko Miyahara
Editors

Quality Assurance in LIS Education

An International and Comparative Study

 Springer

Editors
Makiko Miwa
Center of ICT and Distance Education
The Open University of Japan
Chiba-shi, Japan

Shizuko Miyahara
Sagami Women's University
Sagamihara City
Kanagawa, Japan

ISBN 978-1-4614-6494-5   ISBN 978-1-4614-6495-2 (eBook)
DOI 10.1007/978-1-4614-6495-2
Springer New York Heidelberg Dordrecht London

Library of Congress Control Number: 2014948174

Printed on acid-free paper

Springer is part of Springer Science+Business Media (www.springer.com)

IN MEMORY OF MOHD SHARIF BIN MOHD SAAD

Mohd Sharif Mohd Saad, Ph.D., President of the Librarians Association of Malaysia, 55, passed away on December 9, 2013. He spent his undergraduate years at the Universiti Teknologi MARA, Malaysia (UiTM) and obtained his Masters in Librarianship from Syracuse University, New York, USA. He received his Ph.D. from University of Malaya, Malaysia. The Faculty of Information Management, UiTM became his home as an undergraduate and continued to serve as that special place in his heart for the rest of his life.

A prominent Malaysian Librarian, academician, and an ardent activist for library international networking, Mohd Sharif strongly believed that relations with colleagues around the world would bring international perspectives and benefits to the local librarians' community. Malaysian librarians always referred to him for advice on collaborative activities, organizing international events, conference travel packages, and study tours arrangement. His involvement and presence at major international events was always comforting and reassuring to both hosts and guests alike. He was involved in many national and international conferences, seminars, and training programs and his much sought after advice and contributions for several forthcoming seminars and conferences will be sadly missed. The proposed international digital library conference, to be organized by National Library of Malaysia, was yet another major event where he was expected to assume a key role.

In the information literacy (IL) arena, he was the country's resource person not only manifested through his list of writings on IL, but his direct involvement in this specialized field. A local champion like him has enhanced the ability of this part of the region to be on the right track of IL initiatives, and is coming together with desired outcomes given by UNESCO. He had initiated IL programs in various library institutions in Malaysia. It has brought together Library and Information Studies (LIS) students, educators, and public librarians to instill the importance of having the right perception on IL and imparting IL skills to different levels of learners. In the current few years, the Librarians Association of Malaysia has managed to obtain the Malaysian government-linked corporations to support the IL programs through direct funding and other meaningful contributions. Beside his academic work, Mohd Sharif has a very cordial relationship with LIS leaders and educators not only in Southeast Asia but also from many parts of the world. His ability to bring himself in various communities connected him with people in the industry beyond mere academic confines.

Mohd Sharif's involvement with both practising librarians, national and international, and academia was truly synergistic in his latest role as Deputy Dean, Research and Industrial Linkages, Faculty of Information Management, UiTM. He was consistently hard at work even in the months leading to his passing away. In October, he led the Librarians Association of Malaysia on a mission for an outreach program in Cambodia. Later in the same month, he was in Istanbul, presented a paper for the European Conference for Information Literacy. One day before his passing, he welcomed Prof. Dr. Ursula Georgy from Cologne, Germany to the Faculty of Information Management of UiTM. His absence at the May 2013 CONSAL Executive Board Meeting was clearly noticed and colleagues and buddies like Antonio M. Santos and Dr. Prachark expected his presence at the forthcoming meeting at Chiengmai in May 2014. He was at the A-LIEP 2013 (Asia-Pacific Conference on Library & Information Education & Practice), and presented a paper in July.

The "Application form for potential host countries for RSCAO Mid Term Meeting" was submitted by Mohd Sharif on 1st October 2013 following an invitation by RSCAO through Dr. Dan Dorner, former RSCAO chair, and Dr. Chihfeng P. Lin, the current Chair of RSCAO. Despite Mohd Sharif's demise, the efforts lived on, and the meeting was held in Kuala Lumpur as scheduled. This citation is in no way exhaustive, but in summation Mohd Sharif was a dear friend to many and he is a great loss not only to the profession and the industry but also to the nation as a whole.

He will be greatly missed, and his memory will be cherished forever.

Balqis Suja'
Dr. Rusnah Johare
April 2014

Foreword

This book is published as one of the GlobaLIS (Global Library and Information Science) projects. The purpose of the GlobaLIS project started in 2010 will take an initiative in attaining the globalization of Japanese LIS (Library and Information Science) professional education (see Chap. 2). Main members of the Project are four Japanese scholars; Dr. Miwa (The Open University of Japan), Dr. Kasai (Tamagawa University), Prof. Takeuchi (Chiba University), and Ms. Miyahara (Sagami Women's University) who have wide-ranging experience overseas to share ideas and insights with one another about the LIS educations in the world. Until 2013, we have studied quality assurance and accreditation systems of LIS education in the World focusing on the transparency and reciprocity of professional qualification and academic degrees in LIS. This book is an outcome of the Project.

The purpose of this book is to review and examine the quality assurance systems of LIS education in the specific countries and regions. In the recent progress of globalization, the number of LIS professionals who wish to work overseas has been increasing. However, the standards of the qualifications as an LIS professional are different in each country; therefore the case would be that the qualification of LIS professional is not accepted and authorized in a foreign country.

To be qualified as a professional in a foreign country as well as local, the global standard of quality assurance system is an urgent need for the local LIS degree and/or certification holder. As a reflection of this trend, the great interest in quality assurance system is growing in the study of higher education. However, we can find only few scholars discussed this topic at international conferences and contributed articles to international journal in the LIS study. Although there are some books on comparative LIS education published recently as stated below, no book on LIS quality assurance system has been published until now.

As a specialist of LIS education, therefore, we decided to focus on the quality assurance system of LIS professional education at the university level. This project is the first collective work on LIS quality assurance system which covers various countries, and all authors of the articles are top-level LIS scholars in each country.

We have no doubt that this book has been expected to be published by many LIS scholars, researchers, and students. It is also highly expected those LIS faculties and students use this book as a textbook at graduate/undergraduate LIS program.

This book is divided into three main parts for theoretical and geographical reasons. The first part discusses the background and some theoretical aspect of the recent LIS education. It has three chapters; the first, by Makiko Miwa, GlobaLIS: Toward the Regional Cooperation of Asia-Pacific details the history and current situation of library and information professional education and certification systems on a global level, and it explores opportunities for collaboration in the curriculum development and quality assurance of professional education in the Asia-Pacific region through GlobaLIS (Global Library and Information Science) project.

The second, by Shizuko Miyahara, Regional Quality Assurance System for Higher Education in Southeast Asia examines the current trend and discussions of the quality assurance system in the higher education field in Southeast Asia and, more specifically, the evolution of multilateralism in the region. And the third, by Yumiko Kasai, International Developments in School Library Studies discusses the reforms and reorganization of library and information science study and education with the recent collective work through the Library and Information Professions and Educations Renewal (LIPER) project.

The second part comprises seven chapters which deal with specific countries in Asia-Pacific region with reference to the development of LIS Education and Quality Assurance System. Chapter 4, Japan by Akira Nemoto, describes the current LIS/LISE situations in Japan and discusses how the Japanese experience will be helpful for considering the situations of other Asian and Pacific countries in the twenty-first century.

Chapter 5, China by Li Changqing, discusses the development of LIS education in China and measures LIS education in China qualitatively. Chapter 6, Taiwan by Chihfeng P. Lin, discusses the stages and practices of quality assurance mechanism with reference to the roles of authority, such as the Ministry of Education in Taiwan.

Chapter 7, Malaysia by Mohd Sharif Mohd Saad, Rusnah Johare, and Fuziah Mohd Nadzar introduces the history of LIS education in Malaysia and presents a report on the development of library and Information Studies education, i.e., quality assurance policies and systems implemented in the actual site of university education. Chapter 8, by Lourdes T. David, pointed out that the Philippines has a unique system of ensuring the quality of library education and practice by virtue of the Republic Act 9246 known as the "Philippine Librarianship Act of 2003."

Chapter 9, Thailand by Sujin Butdisuwan, examines the concept of quality assurance in higher education and provides an overview of Library and Information Science (LIS) education and the existing situation of quality assurance systems, guidelines, and standards in Thailand. Chapter 10, Indonesia by Sulistyo-Basuki, L., LIS Education and Quality Assurance System in Asia Pacific: Indonesia reports on varied quality assurance for LIS education in Indonesia.

The third part comprises six chapters which focus on the specific regions in the world. Chapter 11, Europe by Anna Maria Tammaro, defines an influence of globalization to the quality assurance in LIS education and figures out a theoretical and practical current discussion around the quality of LIS education. Chapter 12, North America by Beverly Lynch, describes a great diversity of institutions and programs of LIS educational accreditation in United States and focuses on the voluntary and self-regulatory nature of accreditation for providing a model to meet accreditation requirements on a more global scale. Chapter 13, South America by Monica, describes the accreditation processes of Library and Information Science (LIS) programs in South America and makes a brief account of the outcomes and lessons learned from the experiences in the selected countries such as Mexico, Colombia, and Costa Rica. Chapter 14, Southeast Asia by Shizuko Miyahara, Diversified LIS Programs in Southeast Asia: Historical Background of LIS Education explores a cause of diversity of LIS education in the area with a historical perspective such as colonial legacy and suggests a regional cooperation to overcome such a diversity among developing countries. Chapter 15, Middle East by Sajjad ur Rehman, focuses on the term Gulf Cooperation Council (GCC) nations and explains how those six nations in the Middle East create an LIS education community.

The chapter writers for the area studies were provided common frames of the development of discussion as well as common questions. The first of these guidelines was that authors should focus on more contemporary developments of LIS education or accreditation after an introduction of brief historical overview in each nation or area. Contemporary developments were then specified as those occurring within the area or beyond the border. As discussed in each chapter, we can observe many aspects and cases of LIS cooperation in the nation or region.

In preparing this book, we called many eminent scholars in this area who represent all regions of the globe. Chapters in this book include papers submitted by those authors who responded enthusiastically to our call. We sincerely thank all contributors of this book. Without them we could not complete our mission.

We received a sad news that one of our coauthor Dr. Mohd Sharif Mohd Saad suddenly passed away due to a heart attack on December 9, 2013. We offer our deepest sympathy and condolence to him and hope this book remind everyone of Dr. Sharif and his eminent works.

Chiba-shi, Japan   Makiko Miwa
Sagamihara City, Kanagawa, Japan   Shizuko Miyahara

Contents

Contributors

Mónica Arakaki Pontifical Catholic University of Peru, Lima, Peru

Sujin Butdisuwan Mahasarakham University, Maha Sarakham, Thailand

Changqing Li Peking University, Tokyo, China

Lourdes T. David Ateneo de Manila University, Quezon City, Philippines

Rusnah Johare Universiti Teknologi MARA, Selangor, Malaysia

Yumiko Kasai Tamagawa University, Tokyo, Japan

Chihfeng P. Lin Shih Hsin University, Taipei, Taiwan

Beverly P. Lynch University of California, Los Angeles, CA, USA

Makiko Miwa The Open University, Chiba-shi, Japan

Shizuko Miyahara Sagami Women's University, Sagamihara City, Kanagawa, Japan

Mohd Sharif Mohd Saad Universiti Teknologi MARA, Selangor, Malaysia

Fuziah Mohd Nadzar Universiti Teknologi MARA, Selangor, Malaysia

Akira Nemoto University of Tokyo, Tokyo, Japan

Sajjad ur Rehman Kuwait University, Kuwait, Kuwait

L. Sulistyo-Basuki University of Indonesia, Depok, Indonesia

Anna Maria Tammaro University of Parma, Parma, Italy

Author (Mohd Sharif Mohd Saad) was deceased at the time of publication.

Part I
Background of the Quality Assurance of the LIS Field

Chapter 1
GlobaLIS: Toward the Regional Cooperation in the Education of Library and Information Professionals in the Asia-Pacific Region

Makiko Miwa

1.1 Introduction

This paper opens with a brief exploration of the history and current situation of library and information professional education and certification systems on a global level, and in Japan, and then explores opportunities for collaboration in the curriculum development and quality assurance of professional education in the Asia-Pacific region, using the GlobaLIS (Global Library and Information Science) project as an example.

1.1.1 Global Trends in Accreditation and Quality Assurance in LIS Professional Education

The regional equivalency of qualifications in the Library and Information Science (LIS) profession is currently a hot issue in Asia and Europe. Historically, the American Library Association (ALA) established an accreditation system for the first LIS professional degree programs in the 1950s, and it currently accredits those programs in the USA, Canada, and Puerto Rico (Burnett & Bonnici, 2006). In the UK, the Library Association and the Institute of Information Sciences were merged into the Chartered Institute of Library and Information Professionals (CILIP) in 2002; this organization currently accredits LIS programs in England, Scotland, Wales, and several European countries (Enser, 2002). In Australia, the Australian Library and Information Association (ALIA) accredits LIS programs (Harvey & Higgins, 2003). In the 1990s, collaboration between North America, England, and Australia was established to attain mutual accreditation on a global level.

M. Miwa (✉)
The Open University of Japan, Chiba-shi, Japan
e-mail: miwamaki@ouj.ac.jp

M. Miwa and S. Miyahara (eds.), *Quality Assurance in LIS Education:*
An International and Comparative Study, DOI 10.1007/978-1-4614-6495-2_1,
© Springer Science+Business Media New York 2015

In mainland Europe, the European Association for Library and Information Education and Research (EUCLID) has been engaged in developing a Europe-wide LIS education program compatible with the Bologna Process, which aims to create a European Higher Education Area by making academic degree standards and quality assurance standards more comparable and compatible throughout Europe (Kajberg & Lorring, 2005). In Southeast Asia, a common quality assurance and accreditation system, based on the North American model, has been proposed by members of the Congress of Southeast Asian Librarians (CONSAL), but the actual implementation is still under discussion at several regional conferences, including the Asia-Pacific Conference on Library & Information Education & Practice (A-LIEP) series (Khoo, Majid, & Chaudhry, 2003).

The Education and Training Section (SET) of the International Federation of Library Associations and Institutions (IFLA) has been engaged in the development of procedures for determining the equivalency of degrees granted and the reciprocity of recognition of job qualifications of a variety of LIS programs all over the world. During a round-table discussion held as part of the offshore meeting of the SET committee in Milan in 2009, representatives of the Association for Library and Information Science Education (ALISE), EUCLID, A-LIEP, and the LIS Special Interests Group from Developing Countries exchanged ideas and opinions on the international guidelines for equivalency and reciprocity of qualifications for LIS professionals on a global scale. Through this discussion, basic agreement was reached regarding the establishment of an international resource center for relevant information on LIS education and the implementation of an outcome-based assessment of the LIS professional qualification system. In general, participants wanted the establishment of assessment criteria and an accreditation procedure that reflected the culture and historical contexts of each country and region.

1.1.2 Restructuring and Quality Assurance of LIS Professional Education in Japan

In Japan, the professional qualification system for librarians has been bifurcated into *shisho* (public librarian) and *shisho-kyouyu* (teacher-librarian). The Japanese Library Act defines the role and requirements for qualified *shisho* (librarian) and *shishoho* (assistant librarian). According to the Library Act, a person is qualified as a *shisho/shishoho* if s/he graduated from a polytechnic, college, or university and completed a training program stipulated by the Ministry of Education, Culture, Sports, Science, and Technology (MEXT). The School Library Act defines the role and requirements for a qualified *shisho-kyouyu*. According to the School Library Act, every school (grades 1–12) is required to have a school library and to employ a *shisho-kyouyu* who is qualified as a teacher and has completed a 10-credit training program stipulated by the MEXT. Regrettably, there is no formal qualification system for information professionals in academic and special libraries.

This section reports on our efforts to restructure the qualification system to create a synthesized training system and quality assurance procedure for all types of libraries, focusing on the Library and Information Professionals and Education Renewal (LIPER) project.

LIPER project. The LIPER project began in April 2003 as a 3-year research project funded by a Grant-in-Aid for Scientific Research. Its aims were to study the history, current situation, and future prospects of LIS education in Japan and overseas to assess the requirement for the possible reform of the Japanese LIS professional education system. Since then, the project has advanced to the third stage, LIPER3. A brief description of each of the LIPER generations follows.

- LIPER1 (April 2003–March 2006): This was the first cross-sectional study of the education and training of information professionals in public, school, academic, and special libraries. The main research activities were conducted by four special teams: an education team identified the current situation of LIS education programs, instructors, and students; a public library team identified skills and knowledge required by public librarians, and ways of acquiring and maintaining them; an academic library team identified skills and knowledge required by academic librarians, and ways of acquiring and maintaining them; and a school library team developed ideal images of school librarians or school media specialists and ways of attaining them. Some additional studies were conducted in medical and law libraries, art documentation, and overseas LIS programs. The major findings were: (1) the structure of Japanese LIS education had remained unchanged for 50 years, and the gap between Japanese and overseas LIS education had been ever-increasing, (2) the curricula and contents of LIS education were not well standardized nor integrated into higher education programs, and very few people who obtained a *shisho* certificate gained employment in the public library sector, (3) new areas of education, including IT skills and user behavior, were sought, and (4) many people sought to obtain LIS education for certification as librarians in public libraries even though employment opportunities for full-time librarians were quite limited (Ueda et al., 2005). The LIPER1 project made two major recommendations for the reform of LIS education: (1) establish an LIS examination for students so that they are able to self-evaluate what they have learned through LIS education and obtain better employment opportunities, and (2) introduce a new standard curriculum for information professional education to emphasize core areas of information organization, information resources and services, information systems and retrieval, management, IT, and a better understanding of user behavior (Table 1.1).

In February 2009, the A-LIEP 2009 was held at the University of Tsukuba with a symposium entitled "Future Perspectives in Globalization of Library and Information Professionals." This was an opportunity to share the findings of LIPER1 and LIPER2 with international counterparts.

Table 1.1 Curriculum for information professionals proposed by the LIPER1 project

Category	Course
Basics of Library and Information Science (LIS)	Foundations of Library and Information Science
	Foundations of Information
	Internship
	Research methods
Information Users	Information behavior
	Training of users
Information Resource Organization	Organization of Information and Resources A
	Organization of Information and Resources B
	Practicum of Information Organization
	Practicum of Special Information Organization A
	Practicum of Special Information Organization B
Information Media	Information Media
	Collection Development
	Special Information Media
Information Services	Information Services
	Practicum of Information Services
Information Systems	Foundations of Library Information Systems
	Information Retrieval
	Practicum of Database Design and Development
	Practicum of Information Retrieval
Management	Foundations of Management
	Knowledge Resources Management
	Practicum of Library and Information Services Planning
Digital Information	Management of Digital Libraries
	Foundations of Digital Content
	Application of Digital Content

- LIPER3 (April 2009–March 2014): The LIPER3 project conducted pilot LIS examinations every year and added new examination questions to the question pool. A three-volume textbook of library and information science, which focuses on the framework of the LIS examination, was published. The effects of a series of LIS examinations on the quality assurance of LIS professional training were evaluated by the special team from JSLIS, and their proposal for full implementation of the LIS examination was discussed by the board of directors.

New standard curriculum for public library professionals. The Basic Act of Education was enacted in 1947. Relevant laws, including the Library Act, were revised in 2006, and the Library Act was revised again in 2008. Major changes concerning the LIS professional qualification system included the provision of a curriculum for the qualification of *shisho* to be taught at colleges and universities as stipulated by the MEXT (shown in Table 1.2), and the stipulation that continuing education was required for those who obtained the qualifications of *shisho/shishoho*.

Table 1.2 Comparison of old and new curriculum for public library professionals

Old curriculum		New curriculum			
Course name	Unit	Category		Course name	Unit
Lecture on Lifelong Learning	1	Required courses	Basic courses	Lecture on Lifelong Learning	2
Introduction to Libraries	2			Introduction to Libraries	2
Lecture on Library Management	1			Lecture on Library & Information Technology	2
Lecture on Library Services	2			Library System and Management	2
Lecture on Information Services	2		Courses on Library Services	Lecture on Basic Library Services	2
Practice of Reference Services	1			Lecture on Information Services	2
Practices of Information Retrieval	1			Lecture on Children's Services	2
Lecture on Library Materials	1			Practice of Information Services	2
Lecture on Specialized Materials	2		Courses on Library & Information resources	Lecture on Library & Information Resources	2
Lecture on Organization of Library Materials	1			Lecture on Information Resource Organization	2
Practice of Organization of Library Materials	2			Practice of Information Resource Organization	2
Lecture on Children's Services	1				
History of Books and Libraries	1	Electives		Lecture on Library Basics	1
Lecture on Special Materials	1			Special Lecture on Library Services	1
Lecture on Communication	1			Special Lecture on Library & Information Resources	1
Lecture on Information Technology	1			History of Books and Libraries	1
Special Lecture on Libraries	1			Lecture on Library Facilities	1
				Practice of General Library Works	1
				Practicum in Libraries	1

The revision strengthened the basis for public library professional training. There is no course aimed at academic and/or special library professionals. Thus, the bifurcated condition of the library professional system has been preserved.

LIS examination. Based on its findings, in 2006 the LIPER1 project proposed the introduction of the LIS examination for quality assurance and outcome-based assessment practices of library professionals for all types of libraries (Miwa et al., 2006). The LIPER2 project initiated the pilot testing of the LIS examination.

For the first 3 years, the pilot LIS examination was conducted as a closed examination in collaboration with some *shisho* training programs and LIS major programs. This was done on a voluntary basis by faculty members who wished to examine whether their teaching outcomes adequately prepared their students to take the pilot examination. The first pilot LIS examination was conducted in the fall of 2007 with 549 candidates. A group of instructors in a variety of LIS areas were commissioned to create pilot examination questions. Some team members carried out publicity work by hosting a series of meetings and making presentations at relevant events. The execution of pilot LIS examinations was commissioned to a not-for-profit organization, but team members attended the examinations and observed the candidates. The second and third pilot LIS examinations were conducted in 2008, with 277 candidates, and in 2009, with 302 candidates. The procedure of creating exam questions and assessing candidates was standardized through these three pilot LIS examinations.

The pilot LIS examination was opened to the public in 2010; anyone who wishes to can pay the examination fee and take the examination in order to assess their level of LIS knowledge and skills, and identify any areas for improvement. Practitioners, as well as current students of LIS, take the exam. Candidates receive their rank and average score in the form of a radar chart so that they are able to compare their skills and knowledge with those of other candidates. Faculty members delivering shisho training programs and LIS major programs, whose students sat the examination, are able to assess their teaching outcomes. The LIS examination Web site was opened at http://www.jslis.jp/kentei/top. Past questions with explanations are shown so that candidates and faculty members are able to prepare for the examination. From 2012, candidates have also received a grade ranking of S, A, B, or C. The top-ranked candidates were awarded with remembrance and their names were published on the Web site, with their agreement.

The introduction of the LIS examination as an outcome-based assessment may help the quality assurance of LIS professional education. However, the intention of the LIPER1 proposal was to synthesize the bifurcated professional qualification system by introducing the LIS examination for students and practitioners seeking any type of library and information professional job. This is because synthesis of library professional qualifications is required to fill the existing gap between the global LIS professional education system and the Japanese one. The current knowledge-based society requires LIS professionals to offer a higher level of professional knowledge and skills than those offered by shisho. Thus, we need to establish an advanced LIS professional qualification by introducing an incremental professional system for all types of libraries.

1.2 Problems and Issues in the Japanese Library Professional System

The LIPER project findings and observations regarding the global trends of accreditation and quality assurance in LIS professional education disclosed the limitations and problem areas in the Japanese education system for library and information professionals. This section provides an overview of the limitations and problems of the Japanese system in attaining collaboration with overseas counterparts.

1.2.1 Public Libraries

Article 2 of Japan's Library Act defines public libraries as those libraries established by municipal governments, while Article 4 of the Library Act determines the responsibilities and qualification requirements for *shisho* and *shishoho*, which, except for *shisho-kyouyu*, are the only library professionals established by law in Japan. This condition is inconsistent with global trends in which qualification systems for library professionals in all types of libraries are synthesized. In addition, the qualifications expected of *shisho* are at college level (including junior colleges), while library professionals in North America, Australia, and England require graduate-level professional education, and the European Bologna Process is shifting library professional training from undergraduate toward graduate level.

One of the surprising findings of the LIPER1 project was the overwhelming number of college graduates (more than 10,000) who obtain the *shisho* qualification every year, even though very few of them (about 30) actually find employment as a library professional in a public library. In fact, some of those who obtained the *shisho* qualification found employment in other types of libraries, including academic, school, and special libraries. Many full-time library workers are recruited by municipal governments through the civil service examination, and are only assigned to a library by chance. They are then expected to move to other sections of municipal government organizations as required, even though they have a *shisho* certificate. Thus, they are not able to work in a library for a long enough period to accumulate experiential knowledge and skills.

One of the main reasons for the limited number of *shisho* employment opportunities in public libraries is the rise of commissioned libraries. More public libraries have outsourced their library services to private finance initiatives (PFI), such as companies in library industries and/or public service corporations, in response to the call for the reform and downsizing of the public servant system. Of the 3,274 public libraries, 347 introduced PFI in 2012 (Yuasa, 2012). In addition, many public libraries employ part-time staff members to replace full-time staff who retire. Although most PFI workers and part-time library employees hold a *shisho* certificate, they are forced to work under poor conditions for low wages. Because opportunities for full-time staff are quite limited, their motivation for acquiring and updating knowledge and advanced skills is low.

1.2.2   Academic Libraries

The lack of a formal qualification system for academic libraries is the fundamental problem for quality assurance of the academic library system. The official regulation to support the existence of academic libraries is the Standard for University Chartering, which stipulates the requirements for all types of higher education institutions (Suzuki, 2007). Article 36 of the SUC requires higher education institutions to have "a library, an infirmary, a study room, and a waiting room for students," and Article 38 defines the function of the library as "to systematically collect books, scientific journals, multimedia materials and other information resources required for education and research, according to the type and size of each department," and to "organize and offer these information resources using adequate systems for information processing and dissemination." Article 38 also requires the library to "have professional and/or full-time workers in order to exhibit its full functions." However the SUC does not define the skills and requirements for professional full-time academic library workers. Nevertheless, national universities select their library workers using the civil service examinations; there is no common selection policy for municipal and private colleges/universities. Eventually, many full-time workers hold a certificate for *shisho*, the national certificate for public library professionals. Those who are recruited as full-time academic library workers in municipal and private colleges/universities are routinely relocated to non-library sections of the institutions, reflecting the established custom that prevents academic library workers from accumulating the skills and knowledge required for ever-evolving academic library services.

Recently, some national universities have outsourced their library services in response to the call for the reform to downsize the public servant system. This introduced a critical situation into the professional services of academic libraries. Some special libraries in academia, such as medical libraries, initiated new professional qualification systems similar to those adopted in North America (Sakai, 2010), which aim to provide a new initiative in developing a new LIS professional system in academic libraries as a whole. There is a serious need to introduce an advanced LIS professional system for academic libraries to ensure higher education institutions offer high-level knowledge and services in response to the needs of a knowledge-based society.

1.2.3   School Libraries

The LIS professional system in school libraries has established the role of *shisho-kyouyu* (teacher-librarian) as stipulated by the School Library Act. The Japanese School Library Act, enacted in 1953, requires that every school (grades 1–12) establishes a school library as a reading and learning information center, and employs a *shisho-kyouyu*, with a supplementary provision that states "the placement of *shisho-kyouyu* is optional for the time being" (Kasai, 2006). Many schools did not employ a *shisho-kyouyu* until 1997, when the supplementary provision was revised to require all schools with 12 or more classes employ a *shisho-kyouyu* by March 31, 2003. Although *shisho-kyouyu* have been employed in most schools since 1997, they are employed primarily as classroom teachers. Thus, their administrative responsibility in the school library operation is supplementary.

According to the findings of the LIPER project, the *shisho-kyouyu* system was not fully functional, even after 1997, because *shisho-kyouyu* are too busy with their main responsibilities of teaching classes. Since 1990, Japan's educational policy has emphasized learning through practical experience, and project-based learning has been represented by "general learning" classes, which encompass inquiry-based learning and information exploration expected to be conducted in school libraries. However, school libraries operate under the supervision of *shisho-kyouyu*, who consider school library administration to be a supplementary task, and are not able to respond fully to the new educational policy. In addition, teachers in Japan are accustomed to teaching classes following the government course (curriculum) guidelines that specify what to teach and how to teach for each of the first 12 classes. The guidelines for the "general learning" course were flexible and mostly entrusted to the teachers, who had difficulty creating their own way of teaching the new course.

The targets of School Library Act are elementary schools (grades 1–6), middle schools (grades 7–9), and high schools (grades 10–12), but the function of school libraries differs among them. School libraries in elementary schools may be well operated by the *shisho-kyouyu*, who view library administration as a supplementary task. However, middle- and high-school libraries require massive information resources of higher quality, and are required to teach information literacy skills. These requirements demand substantial time and skills to be fully operational, and these are beyond the supplemental role of *shisho-kyouyu*. Many public high schools had employed full-time clerical staff responsible for the school library operation for a long time. However, the number of these clerical staff diminished and they were replaced by *shisho-kyouyu* after the revision of the School Library Act in 1997.

In addition to the teacher certificate, only five courses are required in order to become a *shisho-kyouyu* in Japan, which is quite limited in comparison with North America, England, and Australia. In the North American model, common basic knowledge and skills are taught in core courses, and specialized knowledge and skills, including those for school library media specialists, are provided in elective courses at master's level. In the British and Australian models, a graduate-level diploma is required as the starting level for LIS professionals. An important area of knowledge and skills in LIS, but something that is lacking in the current *shisho-kyouyu* education system is information literacy training. Since the 1980s, provision of information literacy training has been required in all types of libraries, including school libraries, in Western countries. However, it has not been introduced in school libraries in Japan, even when informatics was introduced as a required course in elementary, middle, and high schools in the 2000s. Thus, there is little connection between the informatics course and the school library, and the importance of information literacy training in school libraries is not widely recognized by school teachers, including *shisho-kyouyu*.

In short, the current school library system in Japan has the following shortcomings:

- The limitation of *shisho-kyouyu* being the only recognized LIS professionals in the school library system in Japan;
- The absence of higher-level qualification systems for those who have greater professional skills and experience than those required for *shisho-kyouyu*; and
- The limited connection between higher education institutions and LIS professional organizations representing school library professionals.

The lack of knowledge and skills on information literacy training held by *shisho-kyouyu* limits the opportunity for school libraries to provide information literacy training for students.

1.2.4 Research Activities

Research in LIS in Japan has leant toward historical studies, and lacks empirical ones (Miwa & Kando, 2003). This is mainly due to the fact that a majority of older faculty members who delivered professional training courses for *shisho* were retired library personnel who did not have formal research training and did not hold a doctoral degree. *Shisho-kyouyu* required only five LIS-related courses, which is also quite limiting for school libraries. As a result, the Japanese LIS professional system is at a critical point; a majority of public libraries cannot offer the professional services demanded by an ICT-based knowledge society, while school libraries cannot provide adequate information literacy training to prepare students to cope with lifelong learning needs. Recently, young faculty members with graduate-level education, mostly master's degrees, have been recruited to deliver professional training courses. A limited number of LIS faculty members have obtained research skills with Ph.D.-level educational backgrounds.

We believe that empirical research can provide an initiative in a fundamental reform of the current library system in Japan. We need to turn the attention of LIS researchers and practitioners toward overseas trends in LIS education and research, so that they can recognize the limitations of the Japanese LIS professional system.

One of the major indices of globalization in research is the ratio of researchers involved in international collaborative research activities. None of the papers published in the *Journal of the American Society for Information Science and Technology* (*JASIS*) between 1981 and 2005 were coauthored by Japanese and overseas researchers, even though researchers from other Asian countries, including China, Korea, and India, published internationally coauthored papers (Chang, 2009).

A content-analysis study of three major research journals in LIS in Japan from 1970 to 2009 identified that the number of authors affiliated to colleges/universities had increased, the number of papers in information science had gradually decreased while those in library science had increased, and the proportion of empirical research had increased (Sugiuchi et al., 2012). These trends reflect the fact that an increasing number of young faculty members who have research skills are being recruited to deliver professional training courses and LIS major programs.

1.2.5 Professional Systems

In North America, England, and Australia, where accreditation systems for LIS professionals are fully operational, a firm level of professionalism has been established; LIS professional associations have influence over the curriculum development and definition of competency required for LIS professionals through accreditation and outcome-based assessment practices. Regrettably, the Japan Library Association (JLA), the single national professional association for public libraries, is not involved in professional accreditation, nor does it collaborate with international counterparts to accommodate mutual accreditation for globalization of the LIS professional system. The main reasons for a weak library professional system in Japan are that:

- Public libraries are not autonomous organizations, being subject to strong governance from municipal governments;
- No clear definitions have been identified for the competencies required for LIS professionals; and
- The authority of library professionals, in terms of social structure and social norms, is weak.

We consider that the Library Act and the School Library Act provide some structural authority to public and school libraries. The crisis the Japanese library profession faces may have been responsible, at least partially, for the legal authority that limited the activities and services of LIS professionals within the framework defined by these laws. Both the education system and the professional career system should be examined in parallel.

Another source of the low status of the library professional organization is the social system of Japanese society, in which government agencies supersede professional systems. In North America, England, and Australia, LIS professionals receive their qualifications from professional organizations. Thus, once a person obtains professional status, they will be considered as a professional throughout their working life. On the other hand, the Japanese government recognizes public library workers as municipal government officials, rather than public library service specialists, and they are often moved to sections other than public libraries. The same is true of school libraries, where a *shisho-kyouyu* is recognized as a teacher who has the same level of teaching responsibility as other classroom teachers, in addition to overseeing the school library operation. Again, there is no formal professional system for academic and special libraries.

1.3 The GlobaLIS Project

The GlobaLIS project is an initiative aimed at achieving globalization of Japanese LIS professional education by defining the requirements for internationally transparent and compatible educational programs for LIS professionals (Miwa, Kasai, &

Miyahara, 2011). Within the framework of LIPER1 and LIPER2, the international team members studied the history and current situation of quality assurance and accreditation systems of LIS professional education programs worldwide. A considerable gap was found between Japanese and Western systems; in Western systems, firm models of qualification systems for LIS professionals are established and a common curriculum for all types of library and information professionals is provided at graduate level for professional development.

In the GlobaLIS project, we asked three research questions in a stepwise manner:

- Step 1: What are the basic requirements for globalizing Japanese LIS professional education?
- Step 2: What kinds of efforts are required for us to attain global collaboration with the Asia-Pacific region in the quality assurance of LIS education? and
- Step 3: How can we attain global collaboration in the development of a common curriculum framework for LIS education in Asia and the Pacific using school librarian training as an instantiation?

In the initial stage of the project, we reviewed the literature and interviewed people involved in regional and global collaboration for quality assurance and mutual accreditation of LIS education programs to identify the three requirements as answers to the first research question:

- Establishment of an office responsible for preparing and administering mutual accreditation of LIS programs with overseas counterparts;
- Implementation of the LIS examination as a means of establishing an outcome-based assessment of LIS education; and
- Establishment of graduate-level LIS education programs to be mutually exchangeable with overseas counterparts.

For the second research question, we identified three areas requiring attention:

- Improvement of international transparency of the Japanese LIS professional system;
- Comparison of the LIS curriculum content with that of overseas counterparts; and
- Stimulation of interest in global trends in LIS education among Japanese LIS educators.

In response to the third research question, we are developing a model curriculum for school library professionals planned for introduction in 2013. This is reported by Dr. Yumiko Kasai in Chap. 3 of this book.

This section compares the curriculum content of overseas graduate-level LIS education programs as a basis for comparison with Japanese curricula, in order to address the second research question.

1.3.1   Global Comparison of Graduate-Level LIS Programs

This study was conducted to address the second research question, which involves (1) improving the international transparency of the LIS professional system in Japan, and (2) comparing the LIS curriculum content in Japan with that of international counterparts.

1.3.1.1   Methods

1. *Data collection*

We developed a database for data gathered collaboratively on core courses in master's-level professional LIS education programs worldwide. The structure of the database is presented in Table 1.3. After several trials of data input and subsequent adjustments, we added instructions for data input as pop-ups appearing over the cursor position. Figure 1.1 presents the interface for data input located at http://www.globalis-net.com/db/inputs/Input. The database retrieval interface is located at http://www.globalis-net.com/db/searchs/. We invited international LIS educators and professionals to input data on core courses of all master's-level LIS programs worldwide.

One of the project members served as an analyst before the data were uploaded onto the public database; when data were missing or unclear, the collaborator who had entered the data was asked to check and modify them. An example of a record shown as a search result is presented in Fig. 1.2. Ultimately, we accumulated data on 286 courses or modules from 34 programs offered in 18 countries.

2. *Data analysis*

The initial analysis of the data involved classifying the courses into "Main category of the course," based on the "Standard Curriculum for Education of Information Professionals" (Table 1.1) proposed by the LIPER1 project in 2006 (Miwa et al., 2006). When we could not classify a course within this framework, we assigned it to the additional category of "Other".

Because our collaborators had already assigned each course or module in the database to one of the eight categories or "Other," and the assignment had been confirmed by the analyst, we retained this categorization of courses. We first calculated the percentage of courses assigned to each category by program, and then pooled percentages by country and then by region.

3. *Limitations of the study*

Because data input was voluntary, very few collaborators provided usable data; thus, most of the data collection and input was carried out by project members. We used open-access syllabi and course descriptions written in English, Chinese, and

Table 1.3 Structure of the LIS curriculum database

Key for retrieval	Item	Data length	Comment
	ID	8	Sequential YYYY9999 (automatic)
✓	Country	50	ISO3166 Country (pull-down menu)
✓	Language	50	ISO639 Language (pull-down menu)
✓	Name of university	200	Mandatory input
	Name of school	200	Mandatory input
	School URL	100	Mandatory input
	Name of program	200	Mandatory input
	Program URL	100	Optional input
	Name of course	200	Mandatory input
	Course description	1,000	Optional input
✓	Main category of the course	100	Mandatory (pull-down menu)
			Basics of LIS
			Information Users
			Information Resource Organization
			Information Media
			Information Services
			Information Systems
			Information Management
			Digital Information
			Other
	Subcategory of the course	100	Optional (pull-down menu)
			Basics of LIS
			Information Users
			Information Resource Organization
			Information Media
			Information Services
			Information Systems
			Information Management
			Digital Information
	Initial input date	10	YYYY/MM/DD (automatic)
	Final revision date	10	YYYY/MM/DD (automatic)
	Your name	100	Mandatory (not shown in the database)
	Your e-mail address	100	Mandatory (not shown in the database)
	Evidence		Upload a PDF file

Japanese as evidence to support the accuracy of the collected data. LIS professional programs taught in any other languages were not included in this study. In addition, IS professional programs that did not publicly display their course descriptions and/ or syllabi were not recorded. Our findings therefore reflect these limitations of the collected data.

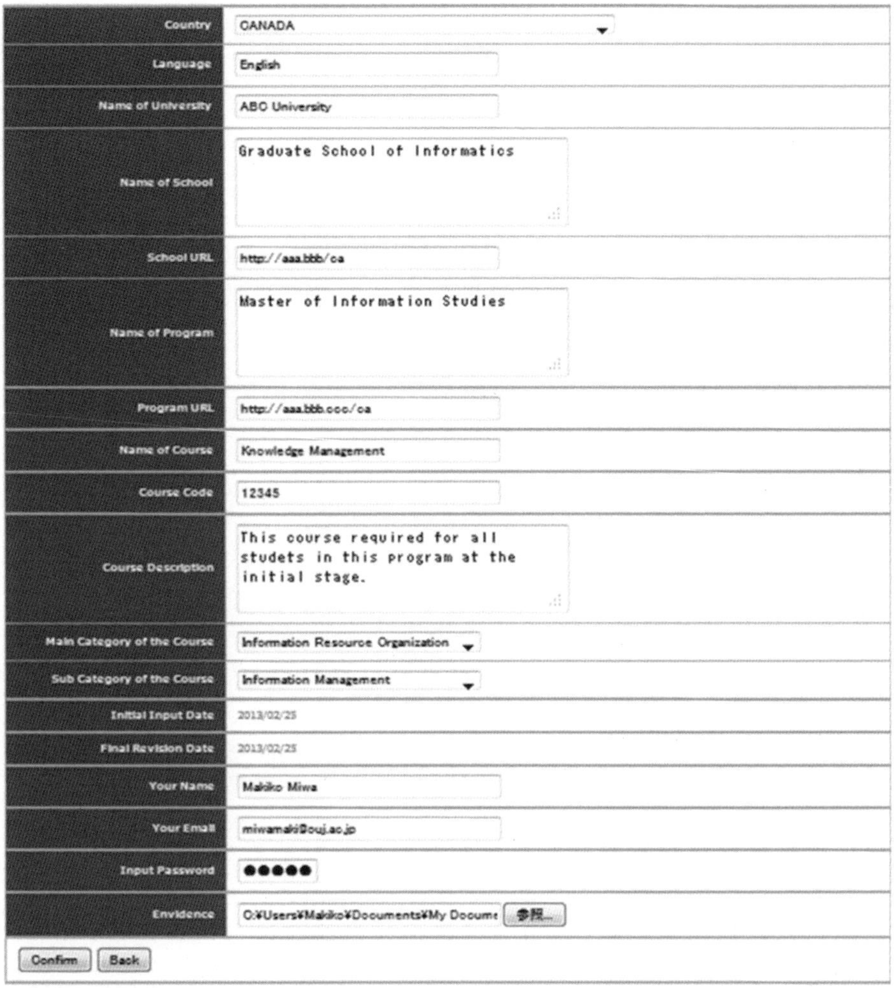

Country	CANADA ▼
Language	English
Name of University	ABC University
Name of School	Graduate School of Informatics
School URL	http://aaa.bbb/ca
Name of Program	Master of Information Studies
Program URL	http://aaa.bbb.ooo/ca
Name of Course	Knowledge Management
Course Code	12345
Course Description	This course required for all studets in this program at the initial stage.
Main Category of the Course	Information Resource Organization ▼
Sub Category of the Course	Information Management ▼
Initial Input Date	2013/02/25
Final Revision Date	2013/02/25
Your Name	Makiko Miwa
Your Email	miwamaki@ouj.ac.jp
Input Password	●●●●●
Envidence	C:¥Users¥Makiko¥Documents¥My Docume [参照...]

Confirm   Back

Fig. 1.1 Data input interface for the LIS curriculum database

1.3.1.2   Results

1. *Characteristics of categories*

Figure 1.3 presents the distribution of courses or modules by category and by region.

As shown in Fig. 1.3, the distribution of the core courses in the LIS programs differs for each region. We now examine the distribution of each category in turn.

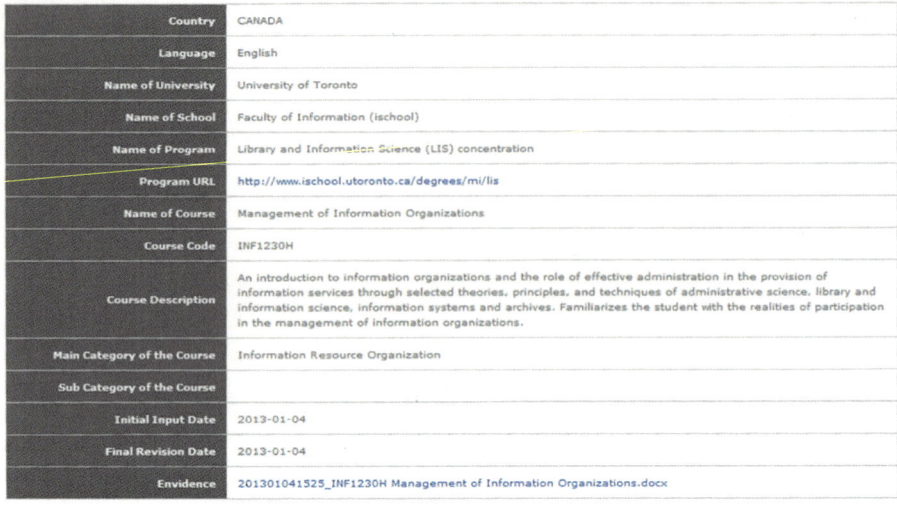

Country	CANADA
Language	English
Name of University	University of Toronto
Name of School	Faculty of Information (ischool)
Name of Program	Library and Information Science (LIS) concentration
Program URL	http://www.ischool.utoronto.ca/degrees/mi/lis
Name of Course	Management of Information Organizations
Course Code	INF1230H
Course Description	An introduction to information organizations and the role of effective administration in the provision of information services through selected theories, principles, and techniques of administrative science, library and information science, information systems and archives. Familiarizes the student with the realities of participation in the management of information organizations.
Main Category of the Course	Information Resource Organization
Sub Category of the Course	
Initial Input Date	2013-01-04
Final Revision Date	2013-01-04
Envidence	201301041525_INF1230H Management of Information Organizations.docx

Fig. 1.2 Sample record from the LIS curriculum database

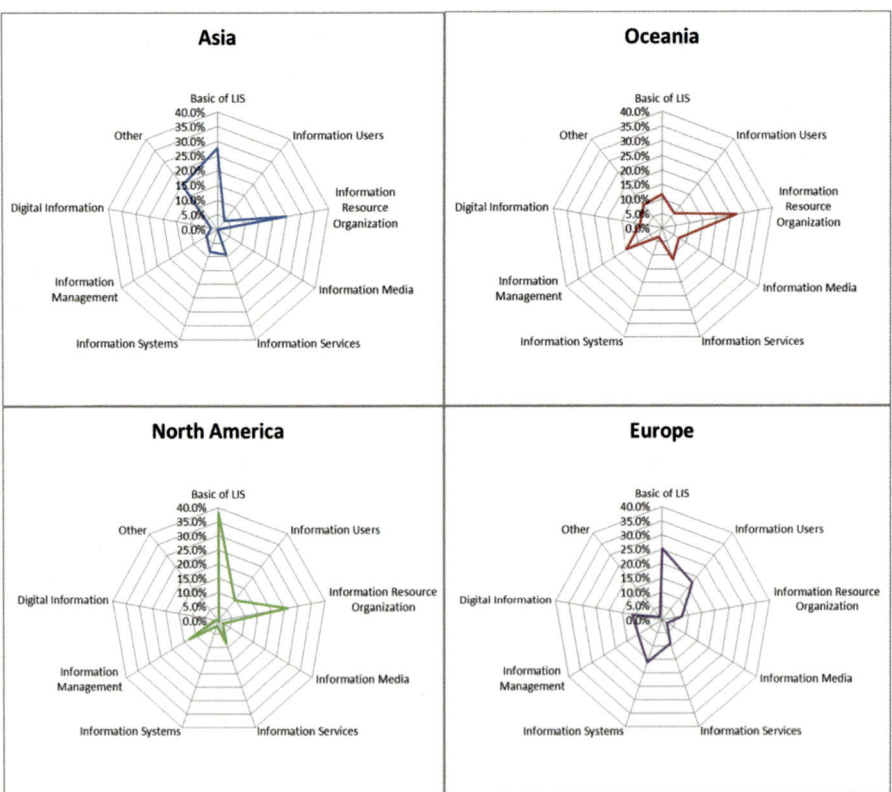

Fig. 1.3 Distribution of courses by category and geographic region

Basics of LIS

Basics of LIS comprises foundation courses such as history, philosophy, and theories of library and/or information science, as well as internship and research methods. In all four regions, more than 10 % of courses fell into this category; the value of 38.0 % for North America reflects the fact that most of the LIS programs in this region have one or more research methods courses as a core course.

Information Users

Information Users includes information behavior and user training. This area of study emerged as a core course in North America in the late 1990s after the central role of users was integrated into the teaching of LIS professionals, as reported in the KALIPER project (Pettigrew & Durrance, 2001). Europe had the largest share of the four regions at 17.5 %. Asia had the smallest share (3.7 %), which may reflect the trend of only minor emphasis on users in LIS education in Asian countries.

Information Resource Organization

Information Resource Organization covers the cataloguing, classification, and summarizing of information objects, that is, the traditional core skills and knowledge in library science. With shares of 24–27 % in Asia, North America, and Oceania, but only 7.4 % in Europe, it appears that *Information Resource Organization* is still considered as the core of LIS in all regions except Europe.

Information Media

Information Media encompasses collection development and a variety of information media. The shares for this core course were low for all four regions, probably because the majority of such courses are offered as electives, as the requisite skills and knowledge are unique to each medium (books, journal articles, music performance, patents, etc.).

Information Services

Information Services covers reference and referral services as well as the delivery of information content. Shares ranged from 8.4 % (North America) to 11.6 % (Oceania). *Information Services* includes a diverse range of topical areas of study, such as medical information, business information, and legal information; these are considered to be electives rather than core courses.

Information Systems

Information Systems includes information retrieval and database and/or Web site design, as well as information architecture. European LIS programs had the largest share at 16.0 %, followed by Asian LIS programs at 8.1 %; the other two regions accounted for just 1.9–3.3 %. These low scores in North America and Oceania may reflect the shift of core courses from system-oriented to user-centered courses, where information seeking rather than information retrieval is included in the core area.

Information Management

Information Management covers knowledge and/or information management, and/ or information service organizations, and general management skills. With information service organizations increasingly demanding greater accountability from information professionals, the emphasis on management skills in LIS programs is growing. As a result, 10–15 % of core courses fell within this category in Europe, North America, and Oceania. Less emphasis was placed on management skills or accountability in the core curriculum in the Asian region (4.5 %).

Digital Information

Digital Information includes the establishment and operation of digital libraries and management of digital content. This category has a relatively high present within core programs in Europe (11.1 %) and Oceania (7.9 %), which may reflect the fact that courses in this category deal with digitization techniques for developing and managing digital heritage materials in response to the increasing use of such materials in libraries, museums, and archives.

Other

Core courses that fell outside the above categories were designated as *Other*. These included a variety of courses such as writing/speech communication skills, foreign language, self-development training, and thesis preparation. The Asian region had the largest share (19.7 %) here. One of the reasons for this was that four programs required students to write a master's thesis before graduating as a library professional.

2. *Characteristics of LIS programs by geographic region*

An examination of the data by geographic region revealed specific characteristics of LIS programs in each region.

Asia

In Asia, the two core courses of *Basics of LIS* and *Information Resource Organization* scored 27.6 % and 24.6 %, respectively. LIS programs in this region generally have an introduction to library (and information) science as the first required course for all students. The slightly larger score for *Information Resource Organization* probably reflects the fact that many Asian LIS programs still emphasize printed books and journals in general studies, rather than digital content and/or materials in a particular genre.

In Oceania, the core course of *Information Resource Organization* accounted for the largest share (27.0 %), which reflects the tradition of technical services in dealing with both printed and digital (networked) information. *Information Management* (14.5 %) and *Information Services* (11.6 %) received greater attention in Oceania than in the other three regions. On the other hand, the relatively low figure of 11.6 % for *Basics of LIS* is something of an anomaly compared with the other regions.

North America

Basics of LIS represented a share of 38.0 % in North America, the largest individual value of all the data, which reflects the fact that most LIS programs in this region have at least one research methods course and one internship or practicum course. However, no course in this region was classified as *Other*, probably because most North American LIS programs are standardized and do not require a thesis or dissertation in order to obtain professional status.

Europe

Compared with the other regions, the core courses of *Information Users* (17.5 %) and Information Systems (16.0 %) stand out in Europe. The former may reflect the European tradition of emphasizing social aspects of information and information-user behavior in LIS programs. The share of 11.1 % for *Digital Information* indicates a growing emphasis on the digitization of cultural heritage materials.

1.3.1.3 Summary

Analyzing the core curricula of LIS professional programs worldwide, we were able to identify the characteristics of programs by category and geographic region. Our discussion is based on the above findings. We identified the following implications, based on our experiential knowledge.

- Even though the data sources were limited, with only 34 programs in four geographic areas of Asia, Oceania, North America, and Europe, general trends emerged from each region.
- In general, the core curriculum content was polarized into traditional print-based library science and newly developed ICT-based i-school curricula.
- In North America, most of the data sources were i-school-based programs, which tend to focus on ICT-oriented library skills where research methods and digital information resource organization are emphasized.
- In Asia, divergence of traditional programs and newly developed i-schools is distinct. Traditional print-oriented library science with thesis requirements still dominates the core curriculum in old programs, while a few newly developed i-schools emphasize ICT-based knowledge management.
- Most of the traditional library science programs in Asia copied the North American curriculum model a long time ago. Some of them have tried to include ICT-oriented courses in order to reflect newly developed ICT markets. However, the shift has not always been successful, mainly because of the limited number of skilled instructors and slow digitization of library materials.
- European programs seem to display trends of integrating libraries, museums, and archives for the digitization of cultural heritage. They stress user behavior and ICT-based skills and knowledge.
- Of the four regions, the curricula of the Oceania region appear to be the most well-balanced. They evenly cover seven areas, and emphasize management skills.

1.3.1.4 Future Study

Because the data collected and analyzed were mostly limited to programs with open-access syllabi and/or course descriptions in English, Chinese, or Japanese, further data are required to gain a holistic picture of worldwide LIS programs. Data collection is thus ongoing with the assistance of instructors drawn from a wider range of programs.

1.3.2 School Library Initiatives for Asia & Pacific (SLAP) Project: Workshop for School Library Practitioners

The SLAP project focuses on the areas of shared interest among practitioners, and selects themes for rescarch and instruction based on the findings of the LIPER and GlobaLIS projects by taking into consideration the current condition of school libraries in the Asia-Pacific region. As part of this project, we held SLAP Forum workshops to provide training for school library practitioners in the Asia-Pacific region. The workshop consisted of three themes; (1) School library and inquiry learning, (2) School library and administration, and (3) School library and curriculum. The specific contents of the SLAP Forum are reported by Dr. Yumiko Kasai in Chap. 3 of this book.

1.3.3 Collection of International and Comparative Studies in LIS Education

This book is one of the outcomes of the GlobaLIS project. The goals of this publication were to cover the three research areas of (1) the quality assurance system of LIS professional education, (2) the transparency and reciprocity of professional qualifications and academic degrees in LIS, and (3) an international comparative study on LIS education. We hope this book has at least partially attained these goals.

1.4 Remaining Issues

The goal of the GlobaLIS project is to take part in an initiative to globalize the training system for Japanese library and information professionals. Many issues must be resolved to attain this goal.

We compared the curriculum contents of LIS professional education in the four regions of Asia, Oceania, North America, and Europe. However, the conclusion of the study is tentative because of the limitations of the data collected. We need to obtain more data, with the assistance of overseas colleagues, to gain a more complete picture of the trends in each of these four regions and to study trends in South America and Africa.

Once we have attained a complete picture of the global trends reflected in the curriculum content of LIS professional training programs, we would like to compare them with the standard Japanese curriculum and share the results and our recommendations with Japanese colleagues and government officials involved in designing professional education programs. In this way, we can help develop a globally compatible LIS professional education system in Japan. To do this, we need to develop competencies required for different types of library and information professionals by identifying the knowledge and skills required by each individual type of library and information professional. We will be able to develop a unique library and information professional training program based on Japanese history and culture. During this process, we need to find ways to implement a qualification system for all types of library and information professionals, rather than one that is limited to current public and school libraries. This will enable us to structure a quality library and information professional system, as recommended by the LIPER project.

Decisions regarding the exchange of credits for LIS professional training programs are currently left to each educational institution in Japan. Currently, the JLA is the only professional association for qualified public library professionals, but it does not collaborate with international counterparts to promote mutual accreditation for globalization and mobility of LIS professionals. We need to establish a new professional organization to develop a professional qualification system for all kinds of library and information professionals, and realize the mutual acceptance of the qualification.

It has been 10 years since the initiation of the LIPER1 project, and we have identified many issues and wrestled with some of them. However, other issues still need to be resolved to reform Japan's library and information professional system; there is a long way to go to attain the goal. We hope to make progress toward the goal by collaborating with overseas colleagues.

References

Burnett, K. M., & Bonnici, L. J. (2006). Contested terrain: Accreditation and the future. Retrieved March 19, 2012, from http://twiki.fe.up.pt/pub/PESIMCI1011/Pesi1011Work2Group2Artigos Seleccionados/Artigo2Grupo2.pdf.

Chang, H. W. (2009). A bibliometric analysis of Asian authorship pattern in JASIS, 1981–2005. *E-Proceedings of the Asia-Pacific Conference on Library & Information Education & Practice*, Tsukuba, Japan, March 6–8. Retrieved January 25, 2013, from http://a-liep.kc.tsukuba.ac.jp/ proceedings/Papers/a31.pdf.

Enser, P. (2002). The role of professional body accreditation in library & information science education in the UK. *Libri, 52,* 214–219. Retrieved March 19, 2012, from http://www.librijournal.org/pdf/2002-4pp214-219.pdf.

Harvey, R., & Higgins, S. (2003). Defining fundamentals and meeting expectations: Trends in LIS education in Australia. *Education for Information, 21,* 149–157. Retrieved March 19, 2012, from http://arizona.openrepository.com/arizona/bitstream/10150/105825/1/S_Higgins_1.pdf.

Kajberg, L., & Lorring, L. (Eds.). (2005). *European curriculum reflections on library and information science education.* Copenhagen, Denmark: The Royal School of Library and Information Science. Retrieved March 19, 2012, from http://dspace-unipr.cilea.it/bitstream/1889/1704/1/ EUCLID%20European%20LIS%20curriculum.pdf.

Kasai, Y. (2006). *School library challenge in Japan—LIPER-SL: Library and information professions and education renewal.* School Library Research Group Report. Retrieved January 25, 2013, from http://www3.ntu.edu.sg/sci/A-LIEP/A-LIEP2006.e-proceedings.htm.

Khoo, C., Majid, S., & Chaudhry, A. S. (2003). Developing an accreditation system for LIS professional education programmes in Southeast Asia: Issues and perspectives. *Malaysian Journal of Library & Information Science, 8*(2), 131–149. Retrieved March 19, 2012, from http://www. ntu.sg/home/assgkhoo/papers/khoo-shaheen-chaudhry.accreditation_for_LIS.2003.pdf.

Miwa, M., & Kando, N. (2003). Japanese LIS research: Content analysis of papers published in recent academic journals. In *Proceedings of the 51st Research Meeting of the Japan Society of Library and Information Science* (pp. 109–112). Ibaragi, Japan: University of Tsukuba.

Miwa, M., Kasai, Y., & Miyahara, S. (2011). GlobaLIS: An effort to describe trends in Japanese LIS education for global collaboration. *Education for Information, 28,* 125–136.

Miwa, M., Ueda, S., Nemoto, A., Oda, M., Nagata, H., & Horikawa, T. (2006). *Final results of the LIPER project in Japan.* 72nd IFLA Council and General Conference, August 20–24, Seoul, Korea. Retrieved January 25, 2013, from http://archive.ifla.org/IV/ifla72/papers/107-Miwa-en.pdf.

Pettigrew, K. E., & Durrance, J. C. (2001). KALIPER project: Final report—KALIPER: Introduction and overview of results. *Journal of Education for Library and Information Science, 42*(39), 170–180.

Sakai, Y. (2010). 5 year anniversary of JMLA Health Sciences Information Professional: An accreditation program of Japan Medical Library Association [in Japanese]. *Journal of Information Processing and Management, 52*(11), 635–644. Retrieved March 17, 2013, from https://www.jstage.jst.go.jp/article/johokanri/52/11/52_11_635/_article/-char/ja/.

Sugiuchi, M., Habu, E., Ueda, S., Kurata, K., Miyata, Y., & Koizumi, M. (2012). The trend of library and information science research in Japan: A content analysis of research articles [in Japanese]. *Library and Information Science, 66,* 127. Retrieved March 13, 2013, from http://www.google.co.jp/url?sa=t&rct=j&q=&esrc=s&source=web&cd=3&ved=0CDgQFjAC&url=http%3A%2F%2Fkoara.lib.keio.ac.jp%2Fxoonips%2Fmodules%2Fxoonips%2Fdownload.php%3Ffile_id%3D68847&ei=oB-mU6qhBIKGkAWmn4DABA&usg=AFQjCNHD-eT4U-r2arxKdzGLl-BvA4kcHcA&bvm=bv.69411363,d.dGI&cad=rja.

Suzuki, N. (2007). The changing role of the university in Japan: The meaning of "cultural mission" in its reform. *19th International Conference on Higher Education,* Babeş-Bolyai University, Cluj Napoca, Romania, 31 August–2 September, 2007. Retrieved March 17, 2013, from www.intconfhighered.org/Naoko%20Suzuki.doc.

Ueda, S., Nemoto, A., Miwa, M., Oda, M., Nagata, H., & Orikawa, T. (2005). LIPER (Library and Information Professions and Education Renewal) Project in Japan. *71st IFLA Council and General Conference, Conference Programme and Proceedings,* August 14–18, 2005. Retrieved March 17, 2013, from http://www.ifla.org/IV/ifla71/papers/051e-Ueda.pdf.

Yuasa, T. (2012). The front line of private finance initiatives: A possible alternative to libraries in the era of decentralization [in Japanese]. Retrieved, from http://2013.libraryfair.jp/node/1254.

Chapter 2
Regional Quality Assurance System for Higher Education in Southeast Asia

Shizuko Miyahara

2.1 Introduction

The higher education population is greater than ever before. The UNESCO Global Education Digest 2009 reported that more than 150 million students belong to higher educational institutions. This number grew by 150 % during the year 2000 and is five times greater than in 1970. Although there is some variation between areas and nations, the higher education population has grown globally in both developed and developing countries (UNESCO Institute for Statistics, 2009).

In addition, the number of overseas students has also been increasing. More than a million students studied at overseas institutions of higher education in the early 1990s, and this number rose to two million in the 2000s. It is expected that by 2025, there will be seven million overseas students (Bohem, Davis, & Pearce, 2002).

However, education beyond borders sometimes causes many problems. The quality of education within a country reflects that country's social system and culture, foreign students are sometimes dissatisfied with the quality of education abroad, and/or some institutions intentionally offer a poor-quality service for foreign students. We still need to discuss whether national regulations or support are absolutely necessary for education or should be a temporary measure during the liberalization of educational service. In the meantime, however, we need regulations and quality assurance measures to avoid the problems of cross-border education directly disadvantaging students.

S. Miyahara (✉)
Sagami Women's University, Sagamihara City, Kanagawa, Japan
e-mail: miyahara_shizuko@isc.sagami-wu.ac.jp

M. Miwa and S. Miyahara (eds.), *Quality Assurance in LIS Education:*
An International and Comparative Study, DOI 10.1007/978-1-4614-6495-2_2,
© Springer Science+Business Media New York 2015

2.2   Higher Education and the General Agreement on Trade in Services

We suggest that one reason for the changes in the higher education market is that the education market has become an object of liberalization through the discussion of the General Agreement on Trade in Services (GATS).

The creation of the GATS was one of the landmark achievements of the Uruguay Round, the results of which came into force in January 1995. The GATS was inspired by essentially the same objectives as its counterpart in merchandise trade, the General Agreement on Tariffs and Trade (GATT): creating a credible and reliable system of international trade rules; ensuring fair and equitable treatment of all participants (principle of nondiscrimination); stimulating economic activity through guaranteed policy bindings; and promoting trade and development through progressive liberalization.

Since the GATS was created in 1995, there has been international pressure towards the liberalization of higher education all over the world. Among other effects, the GATS reinforces commercially driven rationales in the internationalization of higher education and introduces trade rules and disciplines in the regulation of the sector. Higher education is considered to be the global product (Knight, 2003).

2.3   Developing Cross-Border Education

Enrolment in higher education has experienced explosive growth across Asia over the last 20 years, the result of school participation rates, increasing demand of the society and economy for specialised human resources, and the perceived importance of advanced education in subsequent life opportunities (Varghese et al., 2014).

As the number of students who belong to higher educational institutions increase globally, the number of overseas students is also increasing. Until recently, overseas students from Asian countries tended to study at higher education institutions in English-speaking countries such as the UK, Canada, Australia, and New Zealand. Traditionally, overseas study meant that a student in one country would move beyond national boundaries to live and attend university and attain an academic degree or skills (Yamada, 2008).

Currently, the overseas study programs have diversified. New programs are known as cross-border education or transnational education has developed in Western countries. For example, Australia is a leading country in the development of transnational education, which involves the education of students located outside Australia by Australian institutions. In 2012, there were 323,612 international students studying in Australian higher education institutions. Of these, 82,468 were enrolled at campuses outside Australia and a further 25,552 were distance education students. These 108,020 transnational students represent 33.4 % of all higher education international students (Department of Education, Australia, 2014). In the UK, 65.2 % of 135 institutions offer 1,534 transnational programs, and operate 43.6 %of transnational programs in Asia. On the other hand, the USA has focused on managing branch campuses overseas (Drew et al., 2008). In 2009, the USA operated 78 branch campuses, almost half (48 %) of the 162 branch campuses around the world (Becker, 2010).

UNESCO/OECD (2005) defines the meaning of cross-border education as follows:

> Cross-border higher education includes higher education that takes place in situations where the teacher, student, programme, institution/provider or course materials cross national jurisdictional borders. Cross-border higher education may include higher education by public/private and not-for-profit/for-profit providers. It encompasses a wide range of modalities, in a continuum from face-to-face (taking various forms such as students travelling abroad and campuses abroad) to distance learning (using a range of technologies and including e-learning).

Cross-border education can take three forms: (1) student/academic mobility; (2) program mobility; and (3) institution mobility (Knight, 2003; Naidoo, 2006).

Knight noted that cross-border mobility of programmes can be described as "the movement of individual education/training courses and programmes across national borders through face-to-face or distance learning models or a combination thereof" and classified six types of transnational education as follows (Knight, 2007):

1. Franchise
   An arrangement whereby a provider in source Country A authorizes a provider in Country B to deliver its course/programme/service in Country B or other countries. The qualification is awarded by provider in Country A. Arrangements for teaching, management, assessment, profit-sharing, awarding of credit/qualification and so on are customized for each franchise arrangement and must comply with national regulations (if they exist) in Country B.
2. Twinning
   A situation where a provider in source Country A collaborates with a provider located in Country B to develop an articulation system that allows students to take course credits in Country B and/or source Country A. Only one qualification is awarded by provider in source Country A. Arrangements for twinning programmes and awarding of degree usually comply with national regulations of the provider the source Country A.
3. Double/Joint degree
   An arrangement whereby providers in different countries collaborate to offer a programme for which students receive qualifications from both providers, or a joint award from the collaborating partners. Arrangements for programme provision and criteria for awarding the qualifications are customized for each collaborative initiative in accordance with national regulations in each country.
4. Articulation
   Various types of articulation arrangements between providers situated in different countries permit students to gain credit for courses/ programmes offered by all of the collaborating providers. This allows students to gain credit for work done with a provider other than the provider awarding the qualification.
5. Validation
   Validation arrangements between providers in different countries allow Provider B in receiving country to award the qualification of Provider A in source country. In some cases, the source country provider may not offer these courses or awards themselves, which may raise questions about quality.

Table 2.1 Program mobility in higher education of Southeast Asia, 2006 (Macaranas, 2010)

	Cambodia	Malaysia	Laos	Philippines	Indonesia	Singapore	Thailand	Vietnam
Franchise		✓		✓				
Twinning		✓		✓		✓	✓	
Double/Joint Degree		✓		✓	✓	✓	✓	✓
Articulation	✓	✓	✓					
Validation	✓	✓		✓	✓	✓	✓	✓
E-learning/ Distance education	✓	✓		✓	✓	✓	✓	✓

6. Virtual/Distance:

Arrangements where providers deliver courses/programmes to students in different countries through distance and online modes. This may include some face-to-face support for students through domestic study or support centres.

Now, the Asia-Pacific Region stands at the forefront of cross-border education (Yi, 2011). Macaranas (2010) examined the cross-border education managed in Asia. Table 2.1 shows the program mobility in Southeast Asia.

Asian countries that promote higher educational reform began to accept overseas students from both within and outside Asia. In Southeast Asia, the higher education market has expanded rapidly. The competitive situation among higher educational institutions has prompted Asian universities to move aggressively into the region to recruit more students. These transformations of the higher educational market have had substantial effects on the formulation of quality assurance models and qualifications.

2.4 The Meaning of Quality Assurance for Higher Education

International human mobility has a meaningful connection with the quality of universities. If a student with a particular educational background moves to another country, they need to retain the same level of educational quality. Cross-border movement of students and the commodification of higher education require cross-border compatibility, and this in turn requires that the quality of higher education be standardized at a national level.

Accreditation is the main tool used to measure the quality of education. The higher education system of each country has a quality assurance system based on the country's own history and tradition. The accreditation system is considered to be the most effective method of standardizing the quality assurance system of each country (Hata, 2009).

In the USA, the accreditation system uses the unique characteristic of an outcome evaluation system to accept many overseas students. This accreditation system could be globalized to enhance cross-border connection.

In Europe, two programs promote the cross-border mobility of students, faculties, and researchers; ERASMUS aims to establish a network of European universities, and the Bologna Process aims to create a European Higher Education Area (EHEA), based on international cooperation and academic exchange being attractive to European students and staff, as well as those from other parts of the world. Hence, European countries share common values of the quality assurance system, such as the accreditation system in the region (see also Chap. 1 for details).

A vocational certificate is necessary to compare qualifications beyond the border. The knowledge, skills, and ability learned in bachelor's or master's programs should be certified by the educational program, degree, and vocational certificate. Such an outcome evaluation has been trialed in Australia (Horii, 2012).

These trends in the USA and Europe have prompted Asian countries to establish a regional quality assurance system. In this chapter, the development of regional quality assurance systems for higher education in Asia is examined.

2.5 Regional Quality Assurance System for Higher Education in Southeast Asia

As a result of globalization and economic growth, the number of enrolments in higher educational institutions is increasing in Asian countries. The number of overseas students is also increasing. Traditionally Southeast Asian countries have sent excellent students, who are candidates for government leadership, to the USA, the UK, and Australia with official scholarships. Recently, economic growth in Asia has enabled younger students to study in English-speaking countries at their own expense.

In the late 1990s, Australian and UK universities set up branch campuses in Malaysia and Vietnam. For example, four universities from Australia and the UK, including Monash University, opened branch campuses in Malaysia, and a branch campus of the Royal Melbourne Institute of Technology of Australia was opened in Vietnam (Umemiya, 2008).

In addition, the educational programs offered by overseas universities have diversified to include degree programs, franchise-like lectures, twinning programs, double degrees, and connected programs. It is important to consider the meaning of such programs to assure the quality of a degree or a lecture (Kitamura & Sugimura, 2012).

In Singapore and Malaysia, the level of higher education has improved remarkably in recent years. Universities in both countries have accepted many students from Association of Southeast Asian Nations (ASEAN) countries in a positive manner. For example, overseas students at the National University of Singapore made up 22 % of all undergraduate students and 70 % of all graduate students in 2006. In Malaysia, universities have promoted the acceptance of overseas students since the late 1990s. As a result, there are more than 30,000 overseas students at Malaysian universities, and about 8,400 of those are from ASEAN countries.

The Malaysia–Indonesia–Thailand Student Mobility Program, an exchange program, was launched in 2010. This program is a joint governmental project with credit transfers and recognition of credit between universities. Vietnam joined the program in 2012, and the program name was changed to the ASEAN International Mobility for Students Programme (AIMS). In 2013, the Philippines, Brunei, and

Japan also joined. The academic areas, as of 2013, are hospitality and tourism, agriculture, language, culture, international business, food science, technology, engineering, and economics. By September 2011, more than 300 students had joined the program (Umemiya, 2008).

As noted above, educational opportunities for students in Southeast Asian countries have increased. Before the higher educational reform in the 1990s, only some students were able to receive tertiary education. In the 1990s, many Asian countries were able to reform their own higher educational systems in view of the globalization and liberalization of higher education. As a result, the number of students and institutions is increasing. However, the increasing number of higher educational institutions is associated with a decrease in higher education quality, and there has been growing recognition of the necessity to build a regional quality assurance system.

2.6 Regional Quality Assurance System for Higher Education in Asia

In this section, several global or regional quality assurance systems are described.

2.6.1 International Network for Quality Assurance Agencies in Higher Education (INQAAHE)

As noted in the Part II "country report" of this book, a quality assurance system for higher education has already been established in each country. In this article, regional accreditation system trends are examined. At the global level, the International Network for Quality Assurance Agencies in Higher Education (INQAAHE) was established in 1991 as a worldwide association with over 200 organizations active in the theory and practice of quality assurance in higher education.

The purposes of the INQAAHE are to:

1. Create, collect, and disseminate information on current and developing theory and practice in the assessment, improvement, and maintenance of quality in higher education;
2. Undertake or commission research in areas relevant to quality in higher education;
3. Express the collective views of its members on matters relevant to quality in higher education through contacts with international bodies and by other means;
4. Promote the theory and practice of the improvement of quality in higher education;
5. Provide advice and expertise to assist existing and emerging quality assurance agencies;
6. Facilitate links between quality assurance agencies and support networks of quality assurance agencies;
7. Assist members to determine the standards of institutions operating across national borders and facilitate better-informed international recognition of qualifications;

8. Assist in the development and use of credit transfer and credit accumulation schemes to enhance the mobility of students between institutions (within and across national borders);
9. Enable members to be alert to improper quality assurance practices and organizations; and
10. Organize, on request, reviews of the operation of members.

2.6.2 Asia-Pacific Quality Network (APQN)

The regional alliance for quality assurance in Asia, the Asia-Pacific Quality Network (APQN), was launched in 2003. Its mission is "to enhance the quality of higher education in Asia and the Pacific region through strengthening the work of quality assurance agencies and extending the cooperation between them (Asia-Pacific Quality Network)."

A significant project of the APQN was the publication of the UNESCO-APQN Toolkit: Regulating the quality of cross-border education in 2006. This publication complements the Guidelines for quality provision in cross-border higher education published by OECD and UNESCO (UNESCO/OECD, 2005). The Toolkit explains the support plan for the regulation of quality assurance in cross-border education.

There are four main policy objectives in the guidelines:

1. Students/learners should be protected from the risks of misinformation, low-quality provision, and qualifications of limited validity;
2. Qualifications should be readable and transparent in order to increase their international validity and portability. Reliable and user-friendly information sources should facilitate this;
3. Recognition procedures should be transparent, coherent, fair, and reliable and impose as little burden as possible to mobile professionals; and
4. National quality assurance and accreditation agencies need to intensify their international cooperation in order to increase mutual understanding.

2.7 ASEAN University Network (AUN)

The ASEAN University Network (AUN) has promoted regional quality assurance activities in recent years. Since 1998, the AUN has focused on quality assurance for higher education in the region (Table 2.2).

The AUN was established in 1995 with an agreement signed by the ministers of the 10 ASEAN countries responsible for higher education. The academic network initially consisted of the top 17 universities in Southeast Asia. The 4th ASEAN Summit in 1992 called for ASEAN member countries to help hasten the solidarity and development of a regional identity through the promotion of human resource development so as to further strengthen the existing network of leading universities and institutions of higher learning in the region. This led to the establishment of the AUN in November 1995. The original members were 11 universities from six

Table 2.2 Current list of AUN member universities

Country	Name of Institution
Brunei Darussalam	Universiti Brunei Darussalam
Cambodia	Royal University of Phnom Penh
	Royal University of Law and Economics
Indonesia	Universitas Indonesia
	Universitas Gadjah Mada
	Institut Teknologi Bandung
	Universitas Airlangga
Laos	National University of Laos
Malaysia	Universiti Malaya
	Universiti Sains Malaysia
	Universiti Kebangsaan Malaysia
	Universiti Utara Malaysia
	Universiti Putra Malaysia
Myanmar	University of Yangon
	University of Mandalay
	Yangon Institute of Economic
Philippines	Ateneo de Manila University
	De La Salle University
	University of the Philippines
Singapore	National University of Singapore
	Nanyang Technological University
	Singapore Management University
Thailand	Burapha University
	Chulalongkorn University
	Mahidol University
	Chiang Mai University
	Prince of Songkla University
Vietnam	Vietnam National University, Hanoi
	Vietnam National University, Ho Chi Minh City
	Can Tho University

countries. By 2013, this had increased to 21 universities from 10 countries, as shown below (ASEAN University Network, see Table 2.2).

The main objective of the AUN is to strengthen the existing network of cooperation among leading universities in the ASEAN region. This is done by promoting cooperation and solidarity among ASEAN scholars and academics, developing academic and professional human resources, and promoting information dissemination within the ASEAN academic community. There are three types of AUN activities; meetings and conferences, faculty exchanges, and student exchanges (Tan, 2012).

2.7.1 AUN-QA Actual Quality Assessment

Since 2000, the AUN has highlighted the importance of higher education cooperation, particularly "Quality Education toward Quality Assurance" in the ASEAN region. With the collaboration and technical assistance of the European Union (EU),

German Academic Exchange Services (DAAD), German Rectors Conference (HRK), and European Association for Quality Assurance in Higher Education (ENQA), the AUN Quality Assurance (AUN-QA) system was developed and successfully implemented, receiving recognition not only in the ASEAN region and Asia but also in East Africa and Europe.

This system started with quality assessment at program level to ensure the development and production of high-quality graduates in the ASEAN region, particularly in the professional disciplines under the Mutual Recognition Agreement (MRA), to mobilize a qualified workforce across the region to support ASEAN Community Building.

In 1998, it mooted the AUN-QA network, which aimed to develop the QA mechanism to uplift and enhance higher education standards among its members.

With strong and active international collaborations between various partners, the AUN-QA network is fully committed to the continuous development of the Guidelines and Manuals for Institutional Quality Assurance in responding to the needs of ASEAN higher education institutions. There are four quality assurance publications, as outlined in Table 2.3.

The AUN-QA system has now been utilized in more than 40 assessments in the ASEAN region. In this regard, the AUN-QA also actively collaborates with other partners in the region, such as the ASEAN Quality Assurance Network (AQAN) and the SEAMEO Regional Centre for Higher Education and Development (SEAMEO-RIHED) through the signing of the Tripartite Partnership Statement to improve quality assurance in the higher education field. With these successes, the AUN-QA recognized the need to go a step further by developing "Institutional Quality Assurance" in the ASEAN region. The rationale behind this move is that now that most of the higher education institutions in the region have achieved their respective national quality assurance systems, it is time to move beyond the national system and integrate the regional standard.

2.7.2 AUN-QA Actual Quality Assessment at Program Level

AUN-QA started in 1998. The objective is as follows:

> AUN-QA has been fine-tuning the AUN-QA System to support, enhance, and sustain the level of quality assurance practiced by the higher education institutions in ASEAN. This is where QA practices are shared, tested, evaluated, and improved.

On completion of the AUN-QA Actual Assessment, a university qualified with any certification status can use the relevant AUN-QA logo and certificates within

Table 2.3 AUN guidelines and manuals

Publication year	Title of publication
2004	AUN-QA Guidelines
2006	AUN-QA Manual
2009	IAI-QA Training Manual
2011	Guide to AUN Actual Quality Assessment at Programme Level

the validity period. The AUN-QA logo and certificates can only be used on corporate materials, brochures, publicity materials, the Web site, and the premises of the awarded university. The logo and certificate must be used as one entity, and no distortions or modifications are allowed.

In Southeast Asia, the AUN has led quality assurance activities at the regional level. Actual Quality Assessments are expected to be performed 24 times, on 60 programs, by December 2013 (See the Appendix of this chapter).

Umemiya (2008) pointed out that a notable characteristic of the AUN-QA activities is that six of the more advanced ASEAN countries with more experience and resources in quality assurance, namely Singapore, Malaysia, Thailand, Indonesia, the Philippines, and Brunei, have been providing assistance to four of the less developed and experienced partner countries, namely, Cambodia, Laos, Myanmar, and Vietnam.

2.8   Conclusion

This chapter examines the current trends in the regional quality assurance system for higher education in Asia. In conclusion, some problems are clarified.

Each Southeast Asian country has tried to develop a domestic quality assurance mechanism, but the stages of development vary widely. Indonesia, Thailand, the Philippines, Malaysia, and Singapore have already established their quality assurance systems for higher education, while Cambodia, Myanmar, and Laos have yet to complete their systems (Kitamura & Sugimura, 2012). This wide variation is considered to be primarily due to the enrolments in higher education institutions. The enrolment numbers vary quite widely among nations (Fig. 2.1). These differences are barriers to the development of a common framework for quality assurance.

In Southeast Asia, the AUN has implemented regional accreditation at the program level. Although this is still a trial model and has limited participation by the top universities in the region, it is a very meaningful achievement. We need to carefully observe how the situation develops in the future.

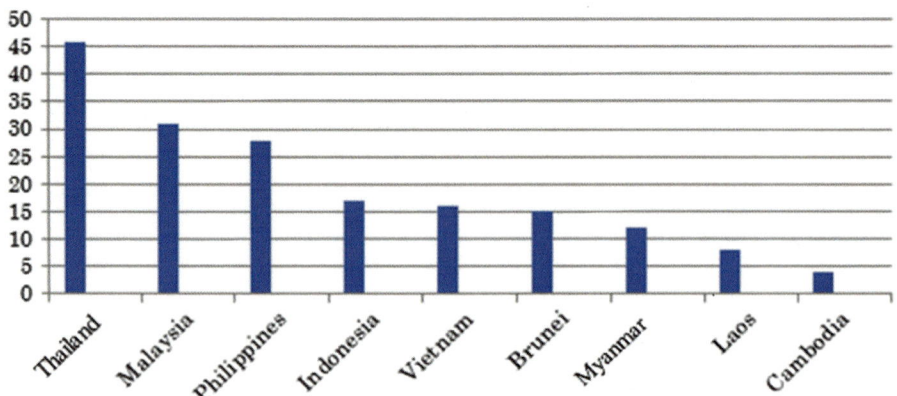

Fig. 2.1 Enrolment ratio for higher educational institutions in Southeast Asia (2005) (UNESCO Institute of Statistics)

Appendix: List of Actual Quality Assessments at Program Level (ASEAN University Network)

Activity	Programs	Host, country	Timing
The 1st AUN Actual Quality Assessment	• Biochemical Engineering • Computer Aided Design and Manufacture Engineering	University of Malaya, Malaysia	12–14 December 2007
The 2nd AUN Actual Quality Assessment	• Applied Economics • Chemical Engineering	De La Salle University, the Philippines	11–13 August 2008
The 3rd AUN Actual Quality Assessment	• Mechanical Engineering • Civil Engineering	Universitas Indonesia, Indonesia	10–12 December 2008
The 4th AUN Actual Quality Assessment	• Physics • Pharmaceutical Science and Technology	Institut Teknologi Bandung, Indonesia	14–17 December 2008
The 5th AUN Actual Quality Assessment	• Pharmaceutical Science • Chemistry • Medical Education	Universitas Gadjah Mada, Indonesia	4–15 October 2009
The 6th AUN Actual Quality Assessment	• Information Technology	Vietnam National University, Hanoi, Vietnam	7–9 December 2009
The 7th AUN Actual Quality Assessment	• Computer Science & Engineering Program • Information Technology Program • Electronic and Telecommunication Program	Vietnam National University, Ho Chi Minh City, Vietnam	10–12 December 2009
The 8th AUN Actual Quality Assessment	• Architecture • Electrical Engineering • Chemical Engineering • Metallurgy and Material Engineering	Universitas Indonesia, Indonesia	12–14 October 2010
The 9th AUN Actual Quality Assessment	• Chemistry • Literature • Psychology	De La Salle University, the Philippines	22–24 November 2010
The 10th AUN Actual Quality Assessment	• International Economics	Vietnam National University, Hanoi, Vietnam	6–8 December 2010
The 11th AUN Actual Quality Assessment	• Biology • Geology Engineering • Civil and Environmental Engineering	Universitas Gadjah Mada, Indonesia	19–21 June 2011

(continued)

(continued)

Activity	Programs	Host, country	Timing
The 12th AUN Actual Quality Assessment	• Bachelor of Science in Applied Corporate Management • Bachelor of Science in Computer Science with Specialization in Software Technology • Bachelor of Science in Physics	De La Salle University, the Philippines	7–9 November 2011
The 13th AUN Actual Quality Assessment	• Biotechnology • Manufacturing Engineering • Vietnamese Studies (University of Technology)	Vietnam National University, Ho Chi Minh City, Vietnam	5–8 December 2011
The 14th AUN Actual Quality Assessment	• Bachelor of Science in Chemistry • Bachelor of Arts in English Teacher Education	Vietnam National University, Hanoi, Vietnam	3–5 May 2012
The 15th AUN Actual Quality Assessment	• Chemistry • Management	Universitas Indonesia, Indonesia	9–11 October 2012
The 16th AUN Actual Quality Assessment	• Bachelor of Business Administration	Vietnam National University, Ho Chi Minh City, Vietnam	12–14 December 2012
The 17th AUN Actual Quality Assessment	• Agronomy and Horticulture • Plant Protection • Aquaculture		
The 18th AUN Actual Quality Assessment	• Department of Agronomy and Horticulture, Faculty of Agriculture • Department of Plant Protection, Faculty of Agriculture • Department of Aquaculture, Faculty of Fisheries and Marine Science	Bogor Agricultural University, Indonesia	22–24 January 2013
The 19th AUN Actual Quality Assessment	• Bachelor of Science in Mathematics • Bachelor of Science in Biology	Vietnam National University, Hanoi, Vietnam	23–25 May 2013
The 20th AUN Actual Quality Assessment	• English • Dentistry • Animal Science and Industry • Legal Science	Universitas Gadja Madah, Indonesia	24–26 October 2013
The 21st AUN Actual Quality Assessment	• Medical Program	Universitas Airlangga, Indonesia	29–31 October 2013

(continued)

(continued)

Activity	Programs	Host, country	Timing
The 22nd AUN Actual Quality Assessment	• Bachelor of Arts in International Studies, Department of International Studies • Bachelor of Science in Mathematics with specialization in Business Application • Bachelor of Science in Statistics, major in Actuarial Science, Department of Mathematics • Bachelor of Science in Civil Engineering, Department of Civil Engineering	De La Salle University, the Philippines	11–13 November 2013
The 23rd AUN Actual Quality Assessment	• Accounting, Faculty of Economics • Industrial Engineering, Faculty of Engineering • Psychology, Faculty of Psychology	Universitas Indonesia, Indonesia	26–28 November 2013
The 24th AUN Actual Quality Assessment	• Master of Arts in English Linguistics • Bachelor of Arts in Linguistics	Vietnam National University, Hanoi, Vietnam	17–19 December 2013

References

Asia-Pacific Quality Network (APQN). Retrieved, from http://www.apqn.org/.

ASEAN University Network. Retrieved, from http://www.aunsec.org/.

ASEAN University Network. *AUN-QA Actual quality assessment at programme level*. Retrieved, from http://www.aunsec.org/programmelevel.php.

Becker, R. (2010). International branch campuses: New trends and directions. *International Higher Education, 58*, 3–4.

Bohem, A., Davis, D., & Pearce, D. (2002). *Global student mobility 2025: Forecasts of the global demand for international higher education*. Australia: IDP Education.

Department of Education, Australia. (2014). *Transnational education in the higher education sector*. Research Snapshot. Retrieved, from https://www.aei.gov.au/research/Research-Snapshots/Transnational%20education_HE_2012.pdf.

Drew, S., et al. (2008). Trans-national education and higher education institutions: Exploring patterns. *Patterns of HE Institutional Activity* (p. 46). Center for Research and Evaluation and Center for Education and Inclusion Research, Sheffield Hallam University. Retrieved, from http://www.shu.ac.uk/_assets/pdf/ceir-TransnationalEducationDIUS-RR-08–07.pdf.

Hata, T. (2009). Sitsu hosyo ni kansusu jyokyo to kadai [Current affairs and agenda of quality quality assurance]. In Hata, T., Yonezawa, A., Sugimoto, K. (Eds.). *Koto kyoiku sitsu hosyono kokusai hikaku.*[International comparative of quality assurance in higher education] (p. 3). Tokyo: Toshindo. [in Japanese].

Horii, Y. (2012). Higashi Asia ken no kyoiku niokeru daigakukan koryu toshitsu hosyo sisutemu [Inter university exchange and quality assurance system of education in East Asia]. In Hayata, Y., & Mochizuki, T. (Eds.), *Daigaku no global-ka to naibu shitsuhosyo* [The globalization of universities and internal quality assurance] (p.81). Kyoto: Koyoshobo. [in Japanese].

INQAAHE. Retrieved, from http://www.inqaahe.org/main/about-inqaahe/constitution/constitution-html.

Kitamura, Y., & Sugimura, M. (2012). *Gekidosuru Asia no daigaku kaikaku. [University reform in exiting Asia]* (p. 11). Tokyo: Sophia University Press [in Japanese].

Knight, J. (2003). *GATS, trade and higher education: Perspective 2003: Where are we?* (p. 30). London, UK: The Observatory on Borderless higher Education.

Knight, J. (2007). Cross-border higher education: Issues and implications for quality assurance and accreditation. In B. C. Sanyal & J. Tres (Eds.), *Higher education in the World 2007. Accreditation for quality assurance: What is at stake* (pp. 134–146). London, UK: Palgrave, Macmillan.

Macaranas, F. M. (2010). Business models in Asia-Pacific transnational education. In C. Findlay & W. G. Tierney (Eds.), *Globalisation and tertiary education in the Asia-Pacific: The changing nature of a dynamic market* (pp. 121–162). Singapore: World Scientific.

Naidoo, V. (2006). International education: A tertiary-level industry update. *Journal of Research in International Education, 5*(3), 323–345.

Tan, K. C. (2012). *The ASEAN University Network Quality Assurance journey: Challenges and lessons learned.* 1st European Network for Engineering Accreditation Annual Conference, 11–12 November, 2012. Retrieved, from http://www.anqahe.org/files/abu_dhabi_2011/FullPapers/Tan_Kay_Shuan Ong_Chee_Bin.pdf.

Umemiya, N. (2008). Tonan Asia niokeru koto kyoiku no shitsu no hosho eno tiikiteki Na torikumi [Regional quality assurance activity in Southeast Asia: Characteristics and driving forces]. *Comparative Education, 37*, 91–11 [in Japanese].

UNESCO/OECD. (2005). *Guidelines for quality provision in cross-border education.* Paris:, UNESCO, OECD. Retrieved, from http://www.unesco.org/education/guidelines_E.indd.pdf.

UNESCO Institute for Statistics. (2009). *Global education digest 2009: Comparing education statistics across the world.* Montreal: UNESCO Institute for Statistics. Retrieved, from http://unesdoc.unesco.org/images/0018/001832/183249e.pdf.

Varghese, N.V., Chiao-Ling C., Montjourides, P., Tran. H., Sigdel, S. Katayama H., Chapman, D. (2014). The reshaping of higher education across Asia. In UNESCO Institute for Statistics (Ed.), *Higher education in Asia: Expanding out, expanding up.* Retrieved, from p://www.uis.unesco.org/Library/Documents/http://www.uis.unesco.org/Library/Documents/higher-education-asia-graduate-university- research-2014-en.pdf.

Yamada, R. (2008). *America no gakusei kakutoku senryaku [Student recruitment strategies institutions American higher education institutions]* (pp. 187–189). Tokyo: Tamagawa University Press [in Japanese].

Yi, C. (2011). Branch campuses in Asia and the Pacific: Definitions, challenges and strategies. *Comparative & International Higher Education, 3*, 8–10.

Chapter 3
International Developments in School Library Studies: A Report on the School Library Initiatives for Asia and Pacific (SLAP) Forum

Yumiko Kasai

3.1 Introduction

The Library and Information Professions and Education Renewal (LIPER) Project, a joint research project made up of members of the Japan Society of Library and Information Science, was modeled on the Kellogg-ALISE Information Professionals and Education Reform (KALIPER) Project announced in 2000 in the USA, which surveyed the state of the shift of librarian training toward information science. Review of librarian training also took place in Britain in 2002, and it is clear that this is a response to the issues of professional training in the information society of the twenty-first century (Fig. 3.1).

The intent of establishing LIPER was "to carry out demonstrative research on issues in librarian training and education, which has not made much progress despite the identification of numerous ideas for improvement over many years, and provide recommendations for its restructuring" in Japanese libraries (LIPER Report, 2006).

Referring to library professionals using the term "information professionals," which in recent years has been used internationally to "denote the profession of librarians in an advanced information society" (LIPER Report, 2006), these recommendations identify a curriculum structure and course system for training such professionals (Figs. 3.2 and 3.3).

Y. Kasai (✉)
Tamagawa University, Tokyo, Japan
e-mail: yumiko_kasai@05.alumni.u-tokyo.ac.jp

M. Miwa and S. Miyahara (eds.), *Quality Assurance in LIS Education:*
An International and Comparative Study, DOI 10.1007/978-1-4614-6495-2_3,
© Springer Science+Business Media New York 2015

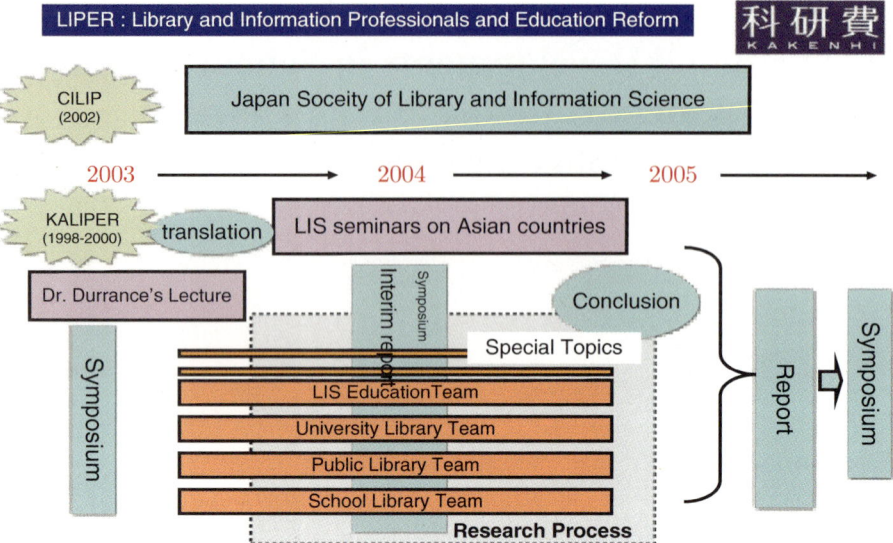

Fig. 3.1 The LIPER Research Framework (from LIPER Report, 2006) (original chart is in Japanese)

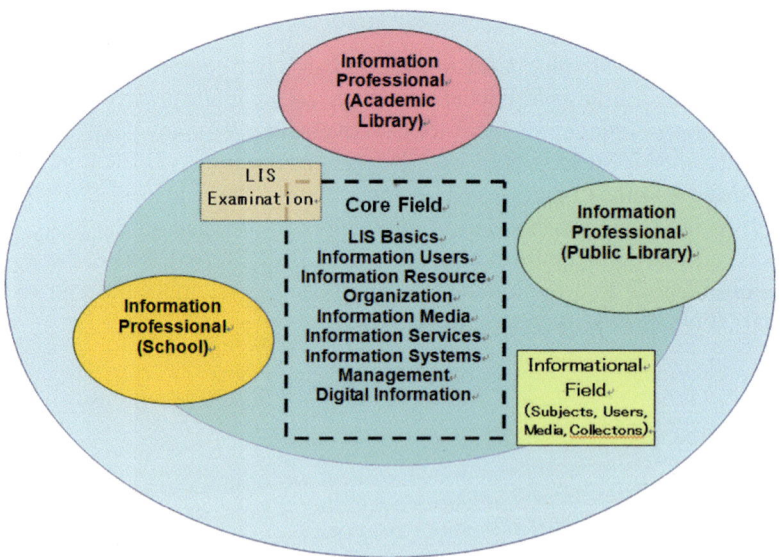

Fig. 3.2 Library and Information Science Curriculum Structure (from LIPER Report, 2006) (original chart is in Japanese)

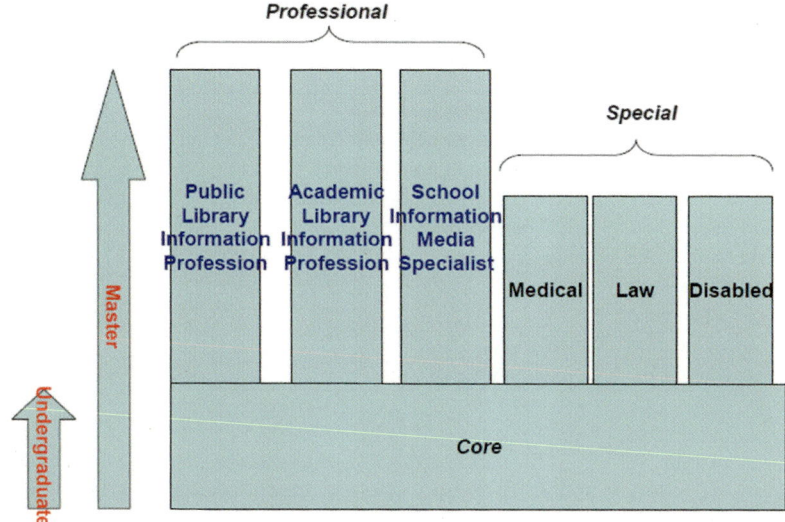

Fig. 3.3 Courses in the Library and Information Science Curriculum (from LIPER Report, 2006) (original chart is in Japanese)

3.2 The GlobaLIS Project

Later, under the succeeding LIPER2 research project, an "International Team" was organized under the theme of "research on international mutual recognition and accreditation of credits in the information professionals training curriculum," as part of "reorganization of library and information science education aiming to train information professionals," And the research was conducted with a focus on "the labor market for library and information science (LIS) professionals in the Asia-Pacific region and trends in international discussion toward quality assurance in LIS professional education" (LIPER2 Report, 2010).

Furthermore, the studies of the International Team achieved independent status as a research project subject to Grants-in-Aid for Scientific Research (Kakenhi), and they were taken over by the Global Library and Information Science (GlobaLIS) Project led by research representative Professor Makiko Miwa of the Open University of Japan. The GlobaLIS Project began in 2010 with the following research agenda:

1. Assessment of equality between the LIPER curriculum (2006) in Japan and world library and information-science curricula
2. Publishing books on international comparative studies concerning library and information-science education
3. Holding workshops for school library practitioners in the Asia-Pacific region

Table 3.1 Subject areas included in the educational programs of European Library and Information Professional Training Institutions (excerpted from Miwa, 2012)

Subject areas
Mediation of cultures in a special European context
The library in the multicultural information society: international and intercultural communication
Cultural heritage and digitization of the cultural heritage
The information society: barriers to the free access to information
Library and society in a historical perspective
Information literacy and learning
Knowledge organization
Knowledge management
Library management and promotion
Information seeking and information retrieval

In studying library and information science educational programs in Europe, Miwa (2012) confirmed that the following subject areas constitute the mainstream in the library and information science education of today (Table 3.1).

3.3 The SLAP Project

Put simply, there are two conceivable points at issue regarding school libraries. The first concerns who should be responsible for management and maintenance of school libraries, and the second is whether professionals in school libraries should be library professionals or education professionals. The situations regarding these points differ even in developed countries such as the USA Britain, and Australia, and are strongly influenced by the circumstances and historical background of each country. However, these also are propositions that cannot be avoided if we are to make any progress in designing a new Asian school library model. Incidentally, the LIPER1 recommendations used the term "information professionals (school)" instead of "(school library)" as used with other types of libraries, to reflect the fact that in the Japanese legal system until now librarians have not necessarily been central in staffing assignments and out of consideration for the choice between education and libraries in future debate.

The SLAP Project, charged with putting the finishing touches on the GlobaLIS Project, began with seeking out areas of shared interest among practitioners and selecting themes for research and instruction, based on the studies conducted in each of the LIPER and GlobaLIS projects and taking into consideration the current conditions of school libraries in the Asia-Pacific region.

3.4 Content of the SLAP Forum

3.4.1 Identifying Themes

The LIPER1 report identified the following eight areas as core areas for information professionals in all library types:

- Library and information science fundamentals
- Information users
- Organization of information resources
- Information media
- Information services
- Information systems
- Management
- Digital information

Among these, the area of "information users" is likely to be deeply related to the school library field as well.

Furthermore, aside from the above core subject areas? common to all library types, the LIPER recommendations also identified the following specialized areas as specialized subjects for "information professionals (school)."

- School education
- Learning information media
- Learning environmental design
- Instruction/learning support
- Children's reading

From these areas, the SLAP Forum workshops to be conducted as training for school library practitioners in the Asia-Pacific region ultimately chose the following three themes: (1) School library and inquiry learning, (2) School library history and administration, and (3) School library and curriculum.

The next sections will report on each theme discussed in the SLAP Forum, held January 12, 2013 at Fukutake Hall of the Graduate School of Interdisciplinary Information Studies, the University of Tokyo, with 30 participants.

3.4.2 Workshop 1: Guided Inquiry

From the 1990s through the 2000s, the biggest proposition in school libraries, chiefly in the USA and the rest of the English-speaking countries, was information literacy. However, in the twenty-first century expectations have focused on not just surviving in an information society but also the image of the ideal human being, possessing the motivation and attitude to continue lifelong learning. At the same time, this can be described as the conclusion derived from the findings of research on humans' information behavior, which until then had been considered straightforwardly. These findings argue that appropriate support and recollection in accordance with each process of handling information are effective, particularly for children and youths, and such education should be embedded into the context of school education.

The Information Search Process (ISP) Model, derived by Carol C. Kuhlthau from analysis of the information behavior of students learning at a school library media center beginning in the 1980s, has provided the theoretical grounding for development of information literacy education worldwide since the 1990s. In recent years Kuhlthau has advocated an inquiry-learning approach, which she calls Guided Inquiry, based on her own ISP Model. Dr. Leslie Maniotes, a coauthor of the recent work Guided Inquiry Design (2012), which examines specific course design for Guided Inquiry, served as the instructor in this workshop.

While Kuhlthau's ISP Model involved analysis from the perspective of the learners who use information, Guided Inquiry focuses on the perspectives of the people providing intervening support and guidance. The design of Guided Inquiry consists of the following eight steps:

- Open
- Immerse
- Explore
- Identify
- Gather
- Create
- Share
- Evaluate

It is a learning approach that provides students with the time and guidance they need to set up their own research themes and to execute the researches (Fig. 3.4).

The workshop included a poster session by Ms. Yoko Noborimoto of the department of information and communication technology (ICT) education at Tamagawa K-12 Academy Tokyo, Japan, who took part in a Guided Inquiry workshop held at Rutgers in June, 2012. This session featured lively questions and debate on policies for putting the Guided Inquiry concept to use in actual class preparations. The workshop also welcomed the unexpected participation of Kuhlthau herself, who had come to Japan to serve as keynote speaker at a international symposium on children's reading to be held the following day in the University of Tokyo, and this session proved to be full of enthusiasm.

3.4.3   Workshop 2: History and Administration of School Libraries

The countries of Asia, which have diverse backgrounds including their own unique languages, cultures, and races, have not developed a model of library professionals that cuts across national borders as in the English-speaking countries and Europe.

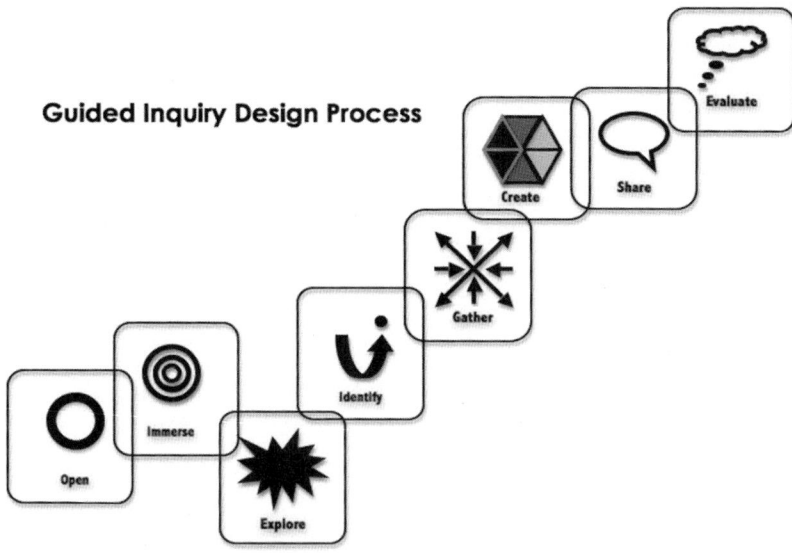

Fig. 3.4 The guided inquiry process (Kuhlthau, Maniotes, & Caspari, 2007)

The Asia-Pacific Conference on Library & Information Education & Practice (A-LIEP), held in Singapore in April 2006, was a momentous international conference in that it expanded the debate over an international certification system for library and information professional educational programs, which until then had been discussed only among the nations of Southeast Asia, to include East Asia, like Japan and Taiwan, as well as the Pacific Rim, including Australia and the USA. Since 2006, A-LIEP has been held every 2 years, and in 2009 the University of Tsukuba in Japan hosted the event. The 2011 meeting in Malaysia featured numerous presentations from Southeast Asian nations and provided an opportunity to deepen our understanding of the circumstances of school libraries in that region.

Dr. Lim Peng Han presented at A-LIEP in Malaysia on historical research on school libraries in Singapore, and that was the opportunity that led to him being asked to lead this workshop.

Dr. Lim first described the current conditions of economic disparities and the information gap in Southeast Asia and then introduced international comparative research on the Knuth school library model, as previous studies (Table 3.2).

Furthermore, on the subject of the current circumstances of school libraries in Asian nations, he conducted a comparison using data from Singapore and Malaysia as representatives of Southeast Asia and from Hong Kong and South Korea as representatives of East Asia (Table 3.3), showing that while each of these countries, like Japan, shows the development of a school library system beginning in the 1950s, the presence and power of school library associations and school library equipment standards are weak and most assigned personnel (denoted as teacher

Table 3.2 Knuth's (1999) international model of school library development

British model (UK and can be applied to developing countries)	American model (USA, Denmark, Canada, and Australia)
1. Education system	
Textbook orientated education systems	School libraries within resource-based education systems
School libraries as book depositories	
A cultural/recreational reading mission	An education mission. School libraries as media centers
2. Staffing of school libraries	
Personnel with inadequate training and role conflict	Staff with dual training
3. Leadership role of School Library Association	
Undeveloped or split professional leadership	Strong professional leadership
4. School Library Services	
Links with public libraries	School libraries within districts
5. School Library Standards	
Underdeveloped professional literature and ineffective standards	Extensive literature and accepted standards
6. Financial and/or statutory government support	

Table 3.3 Comparison of school library development (using Asian countries as an example)

	Singapore	Malaysia	Hong Kong	Korea
Population/per capita income	5.2 million US$56,570	29.0 million US$14,771	7.1 million US$46,291	49.2 million US$ 28,982
Founding of library association	1955	1955	1958	1945
Introduction to modern school librarianship	1955	1955	1963 (grant only)	1950s–1960s
School library association	1969–1980 (none presently)	Nil	HKTLA SLG, HKPTA[a]	NGO (1999)
School library standards	1972 and 1983 (none presently)	Nil	Education dept. (basic standards, 1998)	Master plan (2003–2007)
School library services	Outsource	Nil	Education dept., library section	First master plan (2003–2007)
Ministry of education	Began library development in 1977	Began during the 1960s	Began during the mid-1960s	First master plan (2003–2007)
Teacher librarians	Part-time	Part-time	Full-time	Mostly temporary teacher librarians
ICT infrastructure	IT Masterplan 1997–2002; 2003–2008; 2009–2014	87 Smart schools (1999–2002) 22 % of school libraries had automated system in 2002	Yes	Edunet (2005)

Note: [a]Hong Kong Teacher Librarians Association; School Libraries Group, Hong Kong Professional Teachers' Association

librarians in the table) are not full-time employees. He also shows how outsourcing of school libraries is taking place in Singapore, and in each case there is only an extremely weak foundation for professionals in school libraries.

Dr. Lim's research confirmed that in comparison with the British and American models seen in previous studies, school libraries in Asia are quite far behind, especially in terms of staffing.

Over the long term, it would be desirable to propose development of school library systems suited to the societies and cultures of each of the countries in Asia.

In addition to school library practitioners and researchers from Japan, participants from South Korea, Singapore, and Australia (some of whom were the members of the panel for an international symposium on children's reading held the following day, January 13, at the University of Tokyo) took part in discussions during the workshop, which included lively discussion of each country's school library administration.

While Dr. Lim's session included the raising of propositions directly related to the reasons for holding the SLAP Forum and participants also made some very important comments, of course it would not be possible to identify the path to an Asian school library model from this single session alone, and as such it served as a stimulating first step toward a grand design.

3.4.4   Workshop 3: School Libraries and Curriculum

As mentioned under Sect. 3.4.2 above concerning Workshop 1, in recent years the recognition has become established that learning activities in school libraries are more effective if they are conducted in ways embedded into existing courses. This draws attention to the position of the school library in the curriculum.

Australia's practical research on school libraries is regarded highly worldwide. The author recalls quite vividly hearing in an interview with an American university librarian around 2000, "Today, Australia conducts the best information literacy education."

Over numerous subsequent research visits to Australia, the author had the impression that the field of libraries in Australia is developing very soundly, and that the country was quite adeptly carrying out its own practical library management by adapting the earlier models of Britain and the USA to fit its own circumstances.

Administration in Australia is decentralized among the states, and the state of Victoria, home to Melbourne, which was Australia's largest city from the nineteenth into the early twentieth centuries, occupied a central position in commerce, culture, and education for a long time. The region also played a central role in the development of libraries and the training of librarians as well.

For these reasons, Susan La Marca, who has been active for many years in the School library Association of Victoria, was chosen to lead this seminar, in light of the high regard in which Australia's curriculum is held internationally.

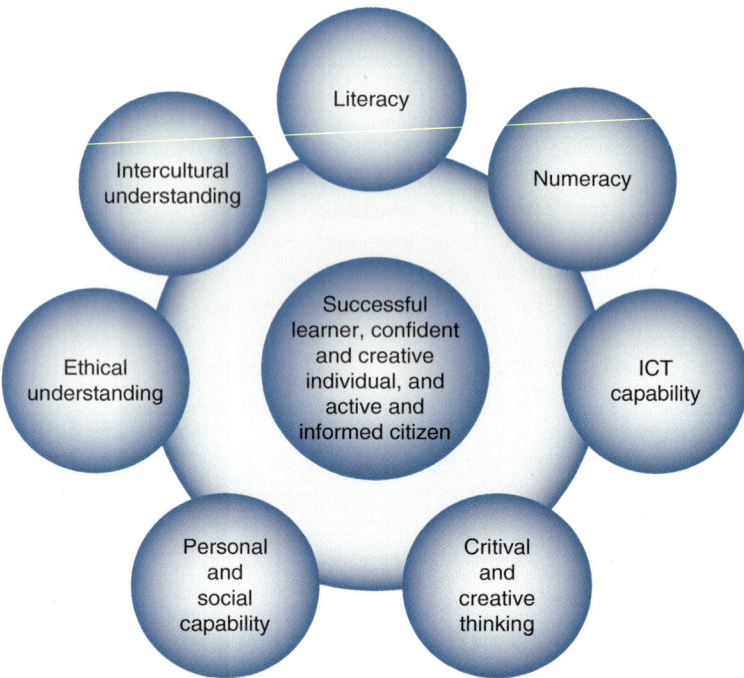

Fig. 3.5 The seven general capabilities of the Australian curriculum (ACARA, 2013)

The Web site of the Australian Curriculum, Assessment, Reporting Authority (ACARA) describes the Australian curriculum.

While there are some differences by state, in Australia education from preschool + 1 through 10th grade is managed together as F-10. Tenth graders are in the final year of secondary school, while 11th and 12th graders, the equivalent of students in their second and third years of senior high school in Japan, make up the senior high school level.

The F-10 curriculum is divided into the four areas of English, mathematics, science, and history, while the following are identified as capability concepts that should be addressed in each course: general capabilities overview, literacy, numeracy, information and communication technology (ICT) capability, critical and creative thinking, personal and social capability, ethical understanding, and intercultural understanding (Fig. 3.5). Each of these seven capability concepts incorporates elements that apply across different courses, and in addition to these the following cross-curriculum priorities are identified as well, suited to Australia's own circumstances: aboriginal and Torres Strait Islander histories and cultures, Asia and Australia's engagement with Asia, and sustainability.

Touching on the interaction between Australia's curriculum and school libraries, Dr. La Marca both described its importance and reported on the diverse efforts to promote reading in which she has been involved over many years. This session was informative and proved highly stimulating to its participants.

3.5 Looking Back on the SLAP Forum

Finally, from the viewpoint of the SLAP Forum's organizer, I conclude this chapter by reveiwing a number of issues.

First, I felt keenly that there still is a long way to go until a standardized school library model for Asia, the issue that SLAP has addressed from the start, will be developed. While recognition of the value of school libraries is increasing in recent years in Japanese society, at the same time staffing issues stand in the way as a crucial impediment to improved school libraries now and into the future. However, through this forum I also was able to recognize that this is an issue common to Asia. It is surmised that factors behind this include sociocultural issues such as the status and roles of experts in society as well as organizational systems and other factors. We should wait for future research to find a solution.

Another issue concerning management of the forum is the fact that there were so few participants from outside Japan. Although we received an inquiry from the Middle East at the application stage, in the end the individual did not participate in the forum. Looking at subjects such as the participants in past A-LIEP meetings, it cannot be denied that it is easier for participants from other Asian countries to attend when the meeting is held in Southeast Asia. Also, since participation from East Asia, particularly from China and South Korea, in international conferences in the library field has been low, there do not seem to be many advantages to holding the meeting in East Asia. It is conceivable that the high prices in Tokyo had a negative effect on participation. Despite this I suggest there is value in the SLAP model and suggest that it be replicated in other Asian countries and regions in the future to the benefit of all of those interested in library professions.

Currently we are preparing to upload a video recording of the SLAP Forum, which was held entirely in English, to a video Web site. We plan to link to the video from the SLAP Web site. I would encourage those who are interested to refer to the site for details.

Biographical Note for SLAP Workshop Facilitators (As of Jan. 2013)

Leslie K. Maniotes, Ph.D., M.Ed., is an educational leader in the Denver Public Schools. A National Board Certified Teacher with 11 years of classroom experience, Maniotes has worked as a Teacher Effectiveness coach and a K-12 literacy specialist in rural and urban Title One schools. She received her doctorate in curriculum and instruction in the content areas from the University of Colorado, Boulder, and master's degree in reading from the University of North Carolina. Recent publication is: Libraries Unlimited's Guided Inquiry: Learning in the 21st Century and Guided Inquiry Design: A Framework for Inquiry in Your School.

Dr. Lim Peng Han has worked in the regional publishing and sports industries from 1980 to 1994. In 1984 he was awarded the "Order of Merit" by the Asian Football Confederation. He began working in academia and academic libraries since 1995. In 2008 he was a Research Fellow at the National Library Board, Singapore. He has written journal articles and conference papers in the history of schooling and school libraries, comparative studies of schooling, publishing and school libraries, sports studies and Southeast Asian Studies.

Dr. Lim is currently the Visiting Research Fellow at the University of Malaya's Sport Centre.

Dr Susan La Marca is a consultant in the areas of YA literature and school libraries, Head of Library at Genazzano FCJ College in Melbourne, the editor of Synergy, for SLAV and associate editor of Viewpoint: on books for young adults. Susan has presented both in the areas of reading culture and school library design and edited six texts in the field of teacher-librarianship including "Rethink: Ideas for Inspiring School Library Design" (SLAV, 2007) and wrote "Designing the Learning Environment" (ACER, 2010). Susan also co-edited "Things a Map Wont Show You: Stories from Australia and Beyond" (Penguin Books, 2012).

Acknowledgements The SLAP Forum was held with the Grant-in-Aid for Scientific Research (Kakenhi) 22300085 and the support of the Japan Society of Library and Information Science. I would like to express my gratitude to all parties involved.

References

ACARA (Australian Curriculum, Assessment, Reporting Authority). (2013). General capabilities in the Australian curriculum. Retrieved July 10, 2014, from http://www.australiancurriculum.edu.au/GeneralCapabilities/Pdf/Overview.

GlobaLIS homepage. Retrieved February 20, 2013, from http://www.globalis-net.com/.

Knuth, R. (1999). On a spectrum: International models of school librarianship. *The Library Quarterly, 69*, 33–56.

Kuhlthau, C., Maniotes, L., & Caspari, A. (2007). *Guided inquiry*. Library Unlimited.

Kuhlthau, C., Maniotes, L., & Caspari, A. (2012). *Guided inquiry design*. Library Unlimited.

LIPER (Library and Information Professional Education Renewal) Report. (2006). Retrieved February 20, 2013, from http://www.jslis.jp/liper/report06/report.htm. (This report was mainly written in Japanese.)

LIPER (Library and Information Professional Education Renewal) 2 Report. (2010). Retrieved February 20, from http://panflute.p.u-tokyo.ac.jp/liper3/?page_id=17. (This report was mainly written in Japanese.)

Miwa, M. (2012). Toward the global harmonization of library and information professionals in Japan. *Johokanri, 54*(10), 611–621. Retrieved February 20, from http://dx.doi.org/10.1241/johokanri.54.611. (This article was written in Japanese.)

School Library Initiatives for Asia and Pacific homepage. Retrieved February 20, from http://www.globalis-net.com/SLAP/.

Ueda, S., Nemoto, A., Miwa, M., Oda, M., Nagata, H., & Horikawa, T. (2005). LIPER (Library and Information Professions and Education Renewal) Project in Japan. *World Library and Information Congress: 71th IFLA General Conference and Council "Libraries—A voyage of discovery"*. Retrieved February 20, from http://www.jslis.jp/liper/report06/ifla-051e.pdf.

Part II
Country Report: LIS Education and Quality Assurance System in Asia-Pacific

Chapter 4
Is the Galapagos Phenomenon Over? Second Consideration of Japanese LIS Education in the International Setting

Akira Nemoto

4.1 Introduction

Until recently, Japanese mobile phone services had been sometimes called a "Galápagos phenomenon" because they quickly evolved in isolation from other countries. Owners used their phones as a television set, e-money machine, train/airplane ticket machine, digital camera, and, of course, a Web browser. A consolidated network terminal provided subscribers with daily information and remote access services in broadcasting, communication, finance, and transportation. However, as many of these services were offered only within Japan, sometimes Japanese mobile phones could not be used outside the country.

But the situation is rapidly changing since early 2010. The so-called smart phones are introduced worldwide, including Japan, where the "Galápagos" phones are cut off from the market. People in Japan are now beginning to use standard smart phones or equivalent machines instead of the old models, some services which cannot be used because the networks often do not support them. Previously, I proposed that the expression "Galápagos phenomenon" can also be applied to Japanese librarianship and Library and Information Science (LIS 図書館情報学) education, in the sense that they were similarly isolated from other countries (Nemoto, 2009). This is a revised version of the previous essay. The Galápagos expression cannot be applied to the recent situation with regard to mobile phones, and we also have to reconsider how relevant it is to LIS education.

In Japan, librarianship was introduced during the early modernization process of the Meiji Period (明治時代1868–1912). The Japanese government and people learned and introduced Western academism, science, and technology. They recognized that libraries and librarianship education were instruments of Western culture to take in and diffuse among the leaders of modernization. But the government

A. Nemoto (✉)
University of Tokyo, Tokyo, Japan
e-mail: anemoto@p.u-tokyo.ac.jp

M. Miwa and S. Miyahara (eds.), *Quality Assurance in LIS Education: An International and Comparative Study*, DOI 10.1007/978-1-4614-6495-2_4, © Springer Science+Business Media New York 2015

prioritized short-term growth of national power and was too engaged in an arms race to support such cultural institutions after the twentieth century began.

Japan reintroduced librarianship during the occupation period of the Allied Forces (連合国軍) just after World War Two. That is why it has been influenced by American librarianship, since the USA was the most powerful country among the Allied Forces. Many Japanese librarians think American librarianship has given us the best model. However, I think Japanese libraries have proceeded in their own way since the occupation ended and they have subsequently selected their own model. What is this model and is it relevant to other countries in Asia and the Pacific region? This is what I will discuss in the following text.

4.2 LIS Education at Japanese Universities

There are five types of LIS education at colleges and universities in Japan. However, in Japan there is no single library act. While the Library Act, which legally establishes that public libraries provide *shisho* (司書), the School Library Act determines that school libraries have *shisho-kyouyu* (司書教諭).

Shisho training—provided by about 200 universities and colleges.

Shisho originally means those who guard and serve book collections; hence, one can earn a librarian certificate in this. Because the Library Act passed in 1950 covers only public libraries, *shisho* was suggested for the public librarian certificate. In general usage, *shisho* is an ambiguous term that is used both for general and public librarians. Students are issued the certificate with at least 24 unit credits of LIS study, which means that 1,080 h of study are required in total, of which 360 h of are included in classes. Most of the study is conducted at an undergraduate or 2-year college level with one or two teaching staff.

Shisho-kyouyu training-provided by more than 100 colleges and universities.

Shisho-kyouyu means "teachers for librarian education certificate" or those who manage and care for school libraries. Students are required to have 10 unit credits in school librarianship in addition to obtaining a teaching certificate. The problems of school libraries and *shisho-kyouyu* are discussed in a later section.

Shisho training programs constitute very small units with one or two faculty, who are comparatively less educated in the LIS career field and sometimes have a career only in librarianship. The officers of the Ministry of Education, Culture, Sports, Science, and Technology (MEXT:文部科学省) check the curriculum and staff on a basic level but they do not have legal authority or specialized knowledge to evaluate them. In fact, the formal training system of *shisho* is rather recent at universities because the article in the Library Act (図書館法) was changed in 2003 to allow colleges and universities to train public librarians. Before this time, all librarians were legally trained at professional sessions managed by universities.

The training conditions of *shisho-kyouyu* are worse than those of *shisho*. The formal instruction of *shisho-kyouyu* now involves only summer training sessions for current teachers and is conducted by universities. Universities can have training sessions for students who are studying for their teaching license but they are legally recognized as training sessions. At schools, *shisho-kyouyu* is concerned with managing school libraries; however, many of them have to teach students all day. Therefore, *gakko-shisho* (学校司書;school librarians) was introduced in schools. These individuals are sometimes irregular or part-time employees. Other programs are not legally regulated and are conducted on a very small scale, except at universities with LIS major programs.

Undergraduate LIS major program—provided by six universities.

The following universities have undergraduate LIS programs:

- Keio University: Library and Information Science, Faculty of Letters
- University of Tsukuba: Knowledge and Library Sciences, School of Informatics
- Surugadai University: Library and Archives Program, Faculty of Media and Information Resources
- Aichi-Shukutoku University: Library and Information Science Series, Faculty of Human Informatics
- Toyo University: Department of Media and Communications, Undergraduate School of Sociology
- Tsurumi University: Department of Library, Archival, and Information Studies, School of Literature

Graduate education (continuing)—provided by eight universities.

Many universities with undergraduate LIS programs also have master's degree programs that are given as continuing LIS programs. Other universities such as Tokyo Gakugei University, Osaka Gakugei University, and Kyushu University provide master's degree programs.

Graduate education (research)—provided by four universities.

Universities with a Ph.D. degree course in LIS are Keio and Tsukuba. In addition, University of Tokyo and Kyoto University are typical academic universities that were set to advance LIS research and cultivate human resources capable of teaching LIS.

Among these, the Keio University academic program is the oldest and started in 1951 as the Japan Library School. It was established to train new librarians following the occupation policies of the GHQ-supported program of the occupation forces, the Allied Powers. After the occupation ended, Keio University absorbed it into their Faculty of Letters program with the financial support provided by the Rockefeller Foundation. They added master's and Ph.D. programs in the 1950s and 1960s. It is recognized as one of the earliest programs to adopt the name LIS in the late 1960s.

The University of Tsukuba program was first set up as a non-formal librarian training program and a small school adjunct to the Imperial Library before World War Two. It became a national junior college of librarianship in the 1960s. In 1979, it was moved to Tsukuba Science City as an independent national university and became the University of Library and Information Science (ULIS). It was one of the largest institutions for training librarians and providing a LIS research program, compared to the Royal School of Library and Information Science in Denmark. In 2004, ULIS was incorporated into the University of Tsukuba, forming a new University of Tsukuba. Now the undergraduate and graduate LIS programs are the largest in Japan with more than 20 faculty members and 300 students.

4.3   A Japanese Model?

One characteristic of Japanese librarianship is its isolation: from other countries, academic disciplines, and even the library profession in general. Librarianship may have also been isolated among academic disciplines in foreign countries but it has been supported by the library profession. However, in Japan, there are no accredited organizations to assure the quality of LIS education.

There are no national standards for LIS education except the *shisho* and *shisho-kyouyu* curricula, which are independent and have minimum standards. There is no national association for LIS education in Japan. Despite this, the Library Science Education Division of the Japan Library Association (JLA 日本図書館協会) and the Japan Society of Library and Information Science (JSLIS 日本図書館情報学会) both have membership systems and have tried to improve library education. However, there are teachers of *shisho* and *shisho-kyouyu* courses who do not belong to these bodies.

We called Japanese LIS education and librarianship a Galápagos phenomenon, but in fact, the situation is not the same as that of mobile phones, where smart phones such as iPhones and Androids began to occupy the market.

In Japan, there are more than 3,000 public libraries staffed with about 15,000 regular status librarians. This includes 7,500 *shisho* staff and more than 1,500 university libraries staffed with about 6,500 regular employees. Many readers may wonder why I do not use the term "professional staff." Generally, we have no professional librarian recruitment programs except for the National Diet Library (NDL 国立国会図書館), some local government public libraries, and a few national universities. "Regular staff" means those who are full-time employees. Many of these regular staff may have studied LIS and might have a *shisho* license, while others may not.

The conditions at school libraries are more complicated. Almost all of Japan's 40,000 public/private schools are equipped with their own school libraries, part of the national requirement according to the School Library Act (学校図書館法1953). Since 2003, those schools having more than 12 classrooms were required to appoint *shisho-kyouyu*, which literally means library-teachers for their school libraries. The license for this is earned by the teachers who studied school librarianship at universities and workshops, for example school library management and school curriculum instruction.

However, *gakko-shisho* had already been assigned as part of a formal or informal requirement at many schools even before it was a legal requirement. *Gakko-shisho* is just a common title that is not legally defined. Due to this condition, some schools are served by *shisho-kyouyu*, others by *gakko-shisho*, some by both, and many by no one.

I explained that we have no formal and integrated professional librarian training programs for university education. The number of new recruitments per year for full-time employment at university and public libraries in Japan is very low, fewer than 100 in total. That is why we are often asked who is working at libraries and who makes book catalogs, since we are managing our libraries with fewer and less-educated library staff.

A set of two hypotheses is provided to explain this situation. The set consists of a highly literate society hypothesis and a generalist bureaucracy hypothesis. These came from my speculation about Japanese modern history and culture.

It could be argued that the Japanese LIS/LISE scene will experience some gradual changes to meet international standards. I also discuss how the Japanese experience will be helpful in considering the situations of other Asian and Pacific countries in the twenty-first century, because in a knowledge-based society, most people might have information literacy and fewer people might come to libraries and ask librarians for assistance.

4.4 Highly Literate Society Hypothesis

Recently, the Edo Period (江戸時代1603–1867) is considered as one of Japanese modernization. Although Japan was under the feudal regime of the Tokugawa Shogunate, literature was developed especially at Edo (now Tokyo), Kyoto, and Osaka during the 250 years of peace throughout Japan. In those years, authors wrote and published novels and dramas. These were sold at bookshops in large cities or by travelling booksellers in rural areas and local cities. This means that the reading public was living both in cities and rural areas. There is some evidence that Japan was one of the most literate countries of the nineteenth century. British sociologist Ronald Dore reported that its school attendance rate was estimated at 40–50 % for boys and 10–15 % for girls in 1968, when the new Meiji Government began (Dore, 1965). He used the earliest Annual Reports of the Ministry of Education (Monbusho 文部省) of the Meiji Government as reference. However, Dore emphasized that this estimate might not have been very precise because the reporting system was not established at that time. He wrote that researchers should explore more precisely what kind of education was offered at schools.

Richard Rubinger of the University of Indiana estimated the illiteracy rate of late nineteenth century Japan and noted that it differed with regard to living areas, classes, and gender between urban and rural areas (Rubinger, 2007). In larger cities, the illiteracy rate was very low (<10 %) but in rural areas it was higher than 50 %. The warrior (samurai) class could generally read because the Han government gave them the opportunity of attending school, called Han-ko (藩校). Rubinger commented about the commoners' engagement in learning (Ibid).

From what has been said so far, it would appear that the Tokugawa regime not only did not fear the common classes' pursuit of learning, it depended heavily upon it, at least insofar as it was limited to the village leadership. Even so, during the seventeenth and much of the eighteenth centuries neither the bakufu nor the domains, with a few prominent exceptions to be taken up later, made significant provisions for the creation and support of schools or other formal institutions that would guarantee quality education for either the samurai or the commoners.

Both male and female commoners engaged in the learning community as the *bakufu* (幕府) governing system was gaining stability. As Dore indicated in his

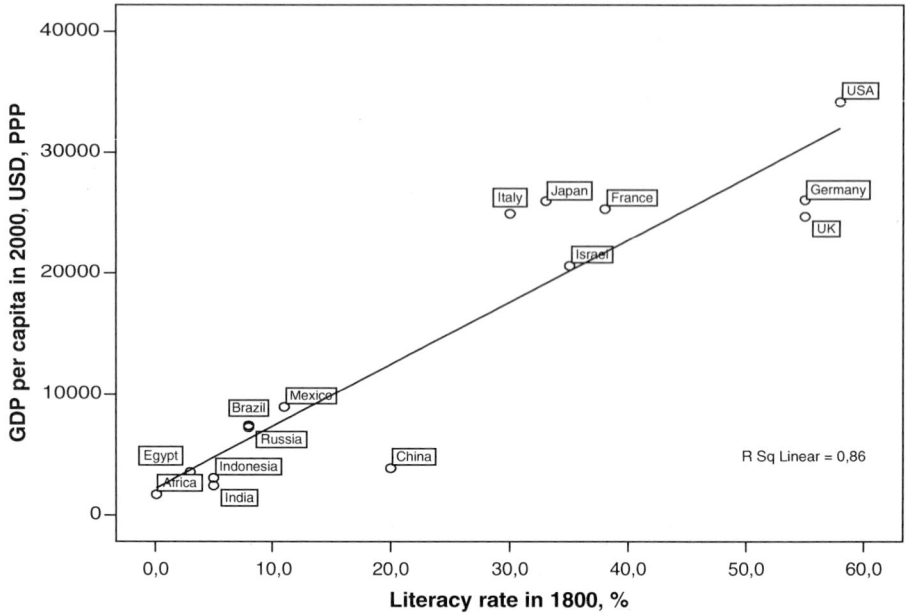

Fig. 4.1 Relationship between literacy and economic development (Korotayev et al., 2006)

report, the conditions would continue to increase the literacy rate at the beginning of the Meiji period. It is difficult to compare Japanese and European literacy development in the same ways. For example, in Europe, couples' signing an autograph at their marriage ceremony was obligatory because it was used to estimate the literacy rate. But in Japan, signing documents was not so rigidly enforced at marriage, childbirth, and military services and it could be easily accomplished by someone else. Therefore, estimating the literacy rate in Japan was difficult and the estimates might have been inaccurate. However, many historians consider people living in Edo and other large cities to have been very literate. In Japanese history, the academic viewpoint that saw the Edo period has changed in the last 20 years from the idea that Meiji was the beginning of modernism to understanding that Edo was actually a period in preparation for modernism. There are many books and articles clarifying the literacy rate during the Edo period. World system theorist Andrey Korotayev and others insisted that the current level of economic wealth relates to the literacy rates in the early nineteenth century in various regions (Korotayev, Malkov, & Khaltourina, 2006). The R^2 coefficient indicates that the correlation between literacy rates in the year 1800 and GDP per capita in 2000 explains 86 % of the entire data dispersion (see Fig. 4.1).

In their discussion, Japan was ranked second in GDP per capita and fifth in literacy rates among 13 countries. This means that Japan was similar to the group of southern European countries, such as France and Italy. As I discussed above, the literacy rates may not be precise in non-Christian areas, but this is a very interesting discovery for understanding the importance of literacy for economic development.

In those days, Japan was geographically located close to the Qing Dynasty (清王朝 1644–1912), one of the historical Chinese dynasties, which maintained their own literate bureaucracies and were major sources of power for building literate societies. This is why the literacy rate has been rather high in the East Asian region recently and its people and societies as a result have been developing economically.

Although Korotayev and others combined literacy and economic development, I prefer to regard literacy as an infrastructure for economic and social development. In addition, literacy was developed not only through reading/writing and calculation training in school-like classrooms (e.g., *terakoya*寺子屋 or *tenaraijo*手習所 in Japan) but also through literary and reading infrastructures. In larger cities such as Edo, Kyoto, Osaka, and Nagoya, there were many publishers and bookshops, suggesting that there were many authors as well as readers. (In Japanese modern history there have been growing interests in Edo period publishing industry. There are many research papers and books in Japanese about the theme, which are omitted because of eliminating the detailed discussions) I think this fact is very important for understanding the role of librarianship in Japanese modernization. Libraries have two main roles in history: one is to transfer older cultural resources from the present to the future, and the other is to deliver mainstream information to mature, mid-, low-, or sub-cultured young or local people. The idea of modern public librarianship began with the national enlightenment movement to create an informed citizenship in the Anglo-Saxon countries. However, in Japan, the second role of creating a reading infrastructure was much needed because people living in cities were already rather literate and the publishing industry had developed to serve their reading needs. For example, in Edo, there were many door-to-door peddlers of rental books who delivered them cheaply. Books were not very expensive and were available in cities and later in local villages. Book markets flourished well, and people enjoyed buying and reading books. This market formed the infrastructure of modern literate society in Japan. Intellectuals inherited the habit of reading from their ancestors, and this habit produced many intellectuals that led and advanced national development in the Meiji period. They sometimes told young people to read books by buying them. The logic behind this is reading a book implies having a dialogue with the author, and to truly understand what the author wrote, readers should underline their favorite sentences and write notes in the margins. It is important for one to possess their own books since their contents reflect and convey an author's idea or message.

Japan was a literate country with a rich literary and cultural tradition when it entered its modernization process. This is one reason why the Japanese government did not consider establishing libraries as a priority during the period of modernization.

4.5   Generalist Bureaucracy Hypothesis

Since the Meiji government began, Japan has maintained its bureaucratic society, in which generalist bureaucrats are dominant within such general organizations as governments and companies. The Meiji Government chose Prussia as its administrative model in the late nineteenth century. In that kingdom/state, high-level bureaucrats made national policies that advanced modernization. This bureaucratic model, which was imitated in Japan, was strong enough to create many modern Japanese institutions, including local governments, companies, and public or private organizations. A century ago, German sociologist Max Weber argued that modern organizations could achieve reasonable and scientific decision-making by adopting bureaucracies (Weber, 1972). A characteristic of bureaucratic organizations is that official business is conducted in strict accordance with the following rules:

1. The duties of each official are delimited by impersonal criteria.
2. The official is given the authority necessary to perform his assigned functions.
3. The means of coercion at his disposal are strictly limited, and the conditions of their use are strictly defined.

Weber also argued that every official's responsibilities and authority are part of a vertical hierarchy, with respective rights of supervision and appeal at each level.

For example, in bureaucracy, those who belong to an organization have their own positions and roles that cannot be changed. Moreover, the rules on how they perform their roles are given from the upper levels of hierarchy and define the limits of power making it possible to act. Of course, there are many exceptions to this theoretical model, but it is thought that organizational employees can often perform best when they are forced to demonstrate their abilities under such rules.

Bureaucracy does not work easily together with professionalism. The action criteria of professionals are found both inside and outside the organization to which they belong. People work to earn their living and serve their community in pursuit of general welfare. This might conflict with the organizational goal. On the other hand, the organizational principle of bureaucracy is to achieve the rational goals of the organization itself, and the criterion lies inside the organization. Administrative departments hire highly capable bureaucrats, including specialists in various areas. These individuals belong to the department as administrative, not professional staff.

Japanese librarians were employed in central or local governments, colleges, universities, schools, or private companies. They also belonged to parent organizations as bureaucrats. The JLA was established in the late nineteenth century, which was the third among library associations internationally. In such organizations, most librarians who were regular employees are now members of the association but they sometimes face contradictions in their identification between bureaucracy and professionalism.

Professionalism does not stand for itself in Japanese librarianship. Since the beginning, Japanese librarians have tried to adopt action plans similar to ALA (American Library Association) or LA (The Library Association, now Chartered Institute of Library and Information Professionals: CILIP). In the time during and directly after the occupation period (1945–1953), the intellectual freedom movement in the USA influenced Japanese librarians in particular. In that time, McCarthyism swept American cities and educational organizations to deny Communism and those who seemed influenced by it. In 1948, ALA members expressed their will to maintain intellectual freedom at libraries by revising the Library Bill of Rights that had been adopted by the members in 1939.

Such actions influenced Japanese librarians, who were asked and sometimes pressured to inform the police about the names of readers of left-wing books in libraries, and to gather and discuss at the JLA conference. In 1954, they adopted the Statement on Intellectual Freedom in Libraries. This was the focus of discussion by professional librarians who wanted to keep their positions ideologically neutral and to be informative for citizens and users. The Statement was slightly revised in 1979 by adding the privacy article. The shorter version of the present statement is as follows: (This is an informal English shorter version of the JLA Statement on Intellectual Freedom in Libraries. It is on the Web page of JLA official sight).[1]

Libraries' most important responsibility is to offer collected materials and facilities to people, whose Right to Know is one of their fundamental human rights. To fulfill their mission, libraries shall recognize the following matters as their proper duties and put them into practice.

- Article 1: Libraries have freedom in collecting their materials.
- Article 2: Libraries secure their freedom of offering their materials.
- Article 3: Libraries guarantee the privacy of users.
- Article 4: Libraries categorically oppose any type of censorship.
- When the freedom of libraries is imperiled, we librarians will work together and devote ourselves to secure it.

This represents the ideal model of actions for professional librarians. However, as noted above, librarians are also organizational staff and sometimes face difficult decisions to judge whether they open, copy or lend certain materials. They might feel a conflict between being professional librarians who make the requested materials accessible to users and being bureaucrats who comply with instructions of the organization to hold back disputable materials from users. Professional librarians sometimes also feel like running away when, for example, they make available any kind of material that indicates controversial aspects of the parent organization.

This bureaucratic model has been maintained for more than 50 years, during which time Japanese librarians have tried to transform it by recognizing their professional roles within the bureaucratic system. However, it is not always easy to combine bureaucracy and professionalism.

4.6 Professionals and Information Technology

Over the past 20 years, Japan has experienced several severe economic depressions. The basic economic idea and policy that Japanese society and the government have chosen is neoliberalism, such as in Anglo-American countries. Neoliberalism is a rational system for evaluating organizations' tasks throughout the process in light of the equality of opportunity, while old liberalism considers

---

[1] http://www.jla.or.jp/portals/0/html/jiyu/english.html.

the equality of results important. The idea behind the former theory involves economic resources to be utilized for advancing the process in each organization, but the idea of the latter involves universal resources to be shared together by society. Therefore, libraries cannot play a role in economic competitions because they are social resources open to everyone.

I would like to argue for a slightly different employment system in Japanese organizations in comparison to other countries. In the late twentieth century, the Japanese economy was strong in that lifetime employment worked better in combination with generalist bureaucracy. Lifetime employment referred to the practice of continuing to employ hired workers up until a fixed retirement age. Under these conditions, temporary staff supported technical and supplementary tasks in the working environment. This is one reason why professionalism was not easy to develop in Japan. Until recently, the view was that librarians are only staff employed to catalogue and classify books.

Information technology has changed this situation by developing means of using computer equipment for improving library services. Widely used library systems and bibliographic utilities have been developed by Information and Communication Technology (ICT) companies and the Machine Readable Cataloguing (MARC) systems, borne out of the cooperation between librarianship and publishing companies. At first, professional librarians contributed to the systems' development but once they were completed and utilized widely, nonprofessional librarians could master these useful systems and electronic resources. Now the tasks are supported by bibliographic utilities and technical supplementary staff who are doing them with assistance from professional staff.

There exists another reason why professional librarians are not easily recognized in bureaucratic hierarchies of Japanese organizations. Although reference and information services are been important aspects of professionalizing Anglo-American libraries, especially after the introduction of bibliographic utility services, they have not been strongly emphasized among Japanese libraries. Both researchers and common citizens believe that information is a private resource to be acquired by themselves, not a public resource accessible to everyone. This is related to the highly literate society hypothesis: users do not trust librarians to be intellectual enough to support them.

4.7 The Japanese Model Revised

Japanese society is highly literate and people can easily obtain their own books in the store. Hence, the Japanese are knowledgeable about books, resulting in less need for professional librarians who can connect books with users. The generalist bureaucracy has not demanded professional librarians, while staff, who may be highly educated and literate but not be professionals or even have a *shisho* license, have functioned as librarians in order to serve users.

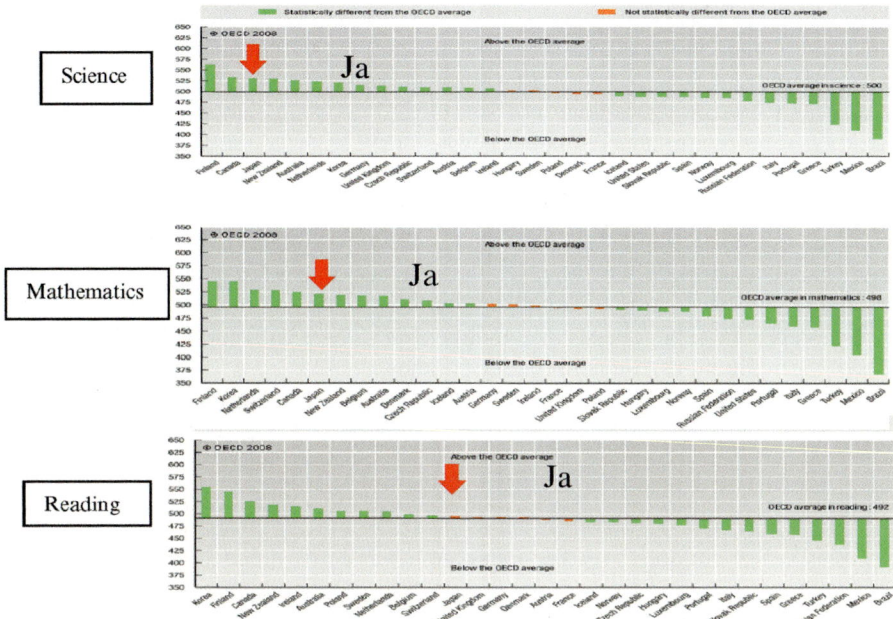

Fig. 4.2   OECD-PISA 2006 scores by nationality

This combination of factors such as literacy, bureaucracy, and ICT has formed the Japanese model. I would regard this as one of the important alternative LIS models in the world. However, we have in fact been forced to modify it recently, especially at the turn of the twenty-first century. One change is that people now read fewer books than before, especially young people. The notion of "literate Japanese" may now be just a myth.

In the 2006 Programme for International Student Assessment (PISA) of the OECD, Japan ranked 12th on the reading literacy test among 26 members, in contrast to ranking 3rd in the science literacy test (see Fig. 4.2).[2] There have been vigorous discussions about these results, e.g., "Reading education lacks a methodology for advancing critical thinking in Japan;" "Young people don't have reading habits;" and "There may be Western cultural bias about definitions and methods of reading literacy tests." However, I think reading literacy has dropped because of structural changes in the media environment, literary culture, and learning expectations over the past 30 years.

---

[2](This data is taken from OECD Factbook 2008: Economic, Environmental and Social Statistics-Outcomes-Education-International   student   assessment,   http://lysander.sourceoecd.org/vl=5906900/cl=29/nw=1/rpsv/factbook/090101.htm).

In 2009, the OECD announced the new PISA results. In the newer edition, the rankings of three kinds of literacies in Japan among 65 countries were 5th in science, 9th in mathematics and 8th in reading. As the reading literacy ranking was higher than 3 years ago among not just OECD countries but all participants, the MEXT and other affiliates felt better because they changed the national curriculum of reading in the Japanese language to solve this problem in the early 2000s.

Students traditionally learned to read the characters and moods in fictional sentences or poems. In those days, reading meant literary reading. In the new curriculum, the main aims of Japanese language are changed from promoting literary reading to communicating with one another by writing and reading factual and scientific sentences. This change in reading materials is influential not only in the school curriculum but in school library policies.

4.8   Reading Promotion and Library Policies

One reason that children's reading promotion was active since the end of the twentieth century was the sense of crisis regarding children's educational conditions relating to the declining birth rate and aging population after the economic bubble collapsed in the early 1990s. There are some ideas and movements for improving these conditions; for example, to improve reading opportunities in communities, change the instructional methodology at schools, and change the level of human investment in school libraries. I would like to discuss the first two tasks first, followed by my evaluation of the third task in the next section.

Although the Japanese Government had often seemed indifferent to the library situation except during the Occupation period, it enacted some policies to promote reading and develop print culture over the past 20 years. In 1993, the MEXT formalized "School Library Materials Five Year Improvement Project," which is one plan for the national taxes allocated to local governments. It also took procedures to deliver 10 billion yen (about 100 million USD) per year for 5 years. These procedures have been continued almost without interruption. The local governments receiving the money provided under the title of school library material costs can also use it for other aims. Thus, all the money was not spent on school library books and media. It is without a doubt that these fiscal measures helped develop public school library collections across the country.

In 2002, the International Library of Children's Literature opened to the public as a branch of the National Diet Library. This was accomplished by reconstructing the old Empire Library Building, built by architect Tadao Ando, at Ueno Park in Tokyo. Construction of the new library seemed to be political because the nonpartisan Federation of Diet Members for Promoting Children and Books was established in 1993 and they supported legislation to establish a new national library so strongly that the decision was made over a very short period. Although I have heard that children's reading is one topic no political party at the Federation of Diet will fight, this means that reading is considered sacred in Japan. Furthermore, political people who believe the highly literate society hypothesis tend to argue for reading from the view of a literacy crisis.

However, a different point of view of literacy states that reading materials include not only traditional and modern literature but also scientific or critical texts and multimedia. The MEXT has been trying to change the national curriculum guidelines that are revised every 10 years to improve reading literacy in line with the international standards. They have introduced "integrated learning" (総合的学習の時間) since the 1990s. In the early 2000s, some critics who insisted that the older curriculum might be better than the newer one attacked the curriculum policies of MEXT. They argued that considering the shortage of weekly school days and class hours, time spent on integrated learning weakened the students' scholastic performances.

The social and political conditions in which school libraries are considered are sometimes loaded with contradictions. School libraries have been regarded more often as learning resource centers than centers of reading materials in the newer curriculum. On the other hand, as young people are expected to have traditional reading literacy at a minimum level, school libraries are regarded as reading resource provision centers. It would be ideal to play these two roles simultaneously but in reality, it is very difficult to select both or either of the two kinds of curriculum policy. Class instructional methodologies for reading are wavering between the traditional and the new.

4.9 LIS Educational Challenges

When the Japan Society of Library and Information Science (JSLIS) held its 50th anniversary ceremony in 2003, it launched a new research project named Library and Information Professionals Education Reform (LIPER) to draw up a blueprint for restructuring LIS education in Japan. The president at that time, Professor Shuichi Ueda, led a research team of some 20 members, including this author, with financial support provided by the Japan Society for the Promotion of Science (JSPS), an independent academic funding organization adjunct to the MEXT.

This team produced a final report (Miwa et al., 2006).

The major findings were the following:

1. The structure of Japanese LIS education has basically remained unchanged for 50 years and the gap between it and LIS education abroad has been steadily increasing.
2. The curricula and contents of LIS education are not well standardized, nor integrated into higher education programs; very few people who obtain a librarian's certificate (*shisho*) procure employment in that field.
3. New areas of education, including IT skills and user behavior, are being sought by librarians.
4. Many people seek an LIS education for certification as librarians even though employment opportunities for full-time librarians are quite limited.

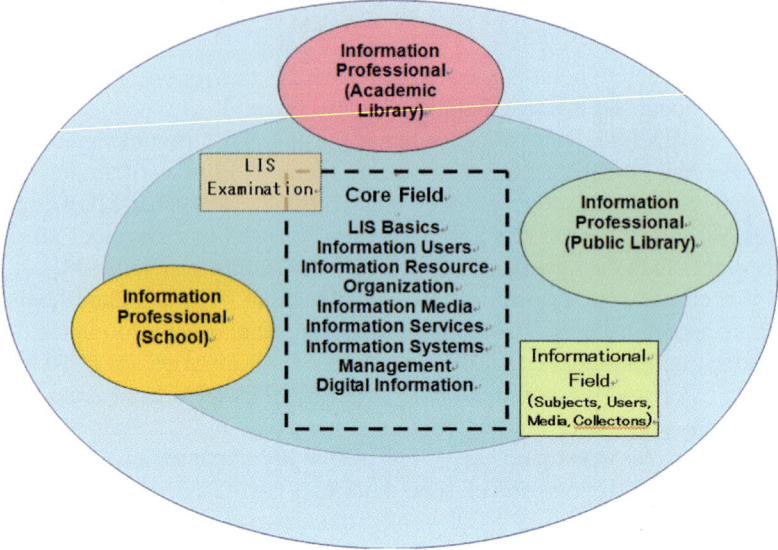

Fig. 4.3 LIPER proposals

5. These findings led the LIPER project to make the following proposals:

   (a) Establish an LIS examination for students so they can self-evaluate what they have learned through LIS education and obtain better employment opportunities.
   (b) Introduce a new standard curriculum for information professional education to emphasize the core areas of information organization, information resources and services, information systems and retrieval, management, IT, and a better understanding of user behavior.

I am afraid that many readers, except perhaps those who grew up in East Asian countries, may wonder why examinations should be introduced. In traditional culture, there was a formal recruiting system based on a written examination administered in a neutral and fair manner that took no account of a person's social status or social class. Although the required abilities differed by society, this kind of written examination for evaluating candidates equally has remained an important part of the modernization process in East Asia.

Of course, this is related to the generalist bureaucracy hypothesis. It may also be one reason why librarians are poorly respected and fewer opportunities exist for evaluating students of LIS education. In order to place librarians and librarianship in the bureaucratic evaluation system, we have been preparing an LIS examination. This might result in a better understanding of the minimum standards for LIS education by examinees and educators.

Figures 4.3 and 4.4 show the curriculum structure of the LIPER proposals. The LIS Examination will be held for the core field of the LIS curriculum, which is expected to be adopted by undergraduate *shisho* training-type courses. We are hoping that thousands of students will take the examination for these courses in the near future.

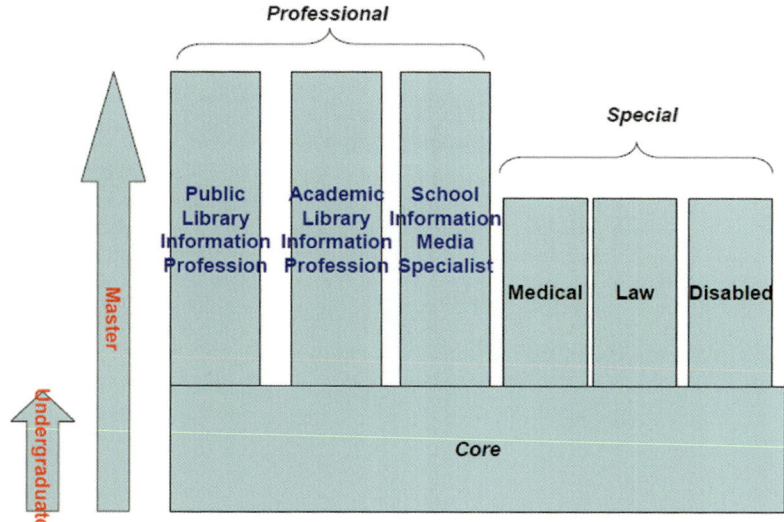

Fig. 4.4 Proposed educational tracks in LIS

Three Information Professional training tracks or a Special Informational training track in Figs. 4.3 and 4.4 will presumably be established within the graduate courses of some universities, which also provide the subjects belonging to the core field. These tracks are examples of further developing LIS education. We have used the term Information Professional to respond flexibly to the changing information environment. As mentioned above, there are now more than five graduate courses in LIS education at Japanese universities. Some universities are likely to undertake new graduate programs for professional librarians and information professionals.

This is currently just a blueprint of our plans. LIPER has been succeeded by the LIPER 2 and LIPER3 Projects of the JSPS research fund since 2006.

4.10 The Past 7 Years

During the 7 years following the final report of LIPER, we have tried to actualize the proposals in LIPER2 and LIPER3. However, we could not launch the second proposal to introduce a new standard curriculum for information professional education because there was no opportunity to do so.

At first, we tried to facilitate discussion among the LIS community by giving the LIS examination. We began to independently prepare an examination with multiple-choice questions and administered this to students in major LIS departments and *shisho* license programs at colleges and universities. We undertook this preparation once a year from 2007 until 2009 and began the formal LIS Examination on the last Sunday of November every year beginning in 2010. In

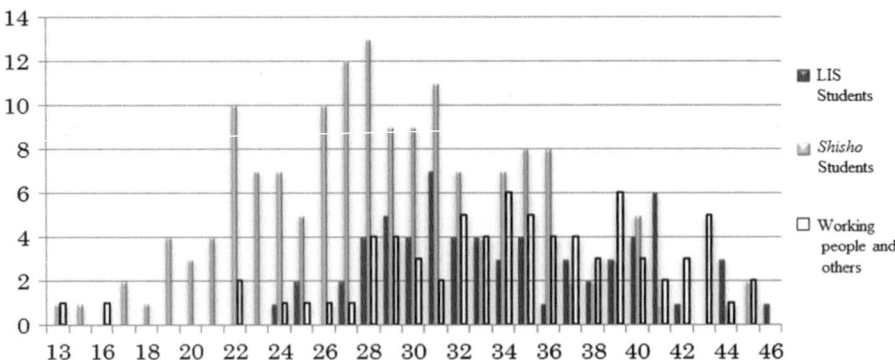

Fig. 4.5 Distribution of exam scores of three groups of examinees

this regard, there are some comparable examinations, such as the Information Retrieval Qualification Examination by the Information Science and Technology Association of Japan (INFOSTA). Our examination is taken by students on the same day as the INFOSTA exam.

I would like to discuss the results of the most recent exam held in November, 2012. The number of examinees was 299 at five open examination places and two closed places for exams. This number may be very low in comparison to the number of students taking *shisho* license courses throughout Japan, which is more than 10,000. However, this is a self-evaluative test and does not give any assurance of obtaining a legal license.

There are three groups of examinees: students having an LIS major, students in *shisho* courses, and working people. The last category consists of professional librarians, contract workers employed as librarians and other working people. The proportion of LIS majors is 21.5 % and their average score is 34.4 (out of a total of 50). The *shisho* students are similarly represented: 53.5 % and 28.9, and working people 24.9 % and 34.4. These data are shown in Fig. 4.5. We can see that the distributions of LIS majors and working people are similar and the *shisho* course students are more numerous but have lower test scores.

The examination consists of eight subject fields. Distributions of the three examinee groups in each field are shown in Fig. 4.6. Apparently, there are differences between the combinations of LIS students, working people and *shisho* course students in the fields of information organization, information media, information systems and digital information. There are minor differences between the three groups in the fields of LIS basics, information users, information services and management and administration. In the figures, we can easily notice weak points in the technical fields of the curriculum for *shisho* courses. In 2012, a new legal curriculum for *shisho* courses was introduced and all the students must take a new course in "library and information technology." We are keen to observe how this change will influence the examination scores.

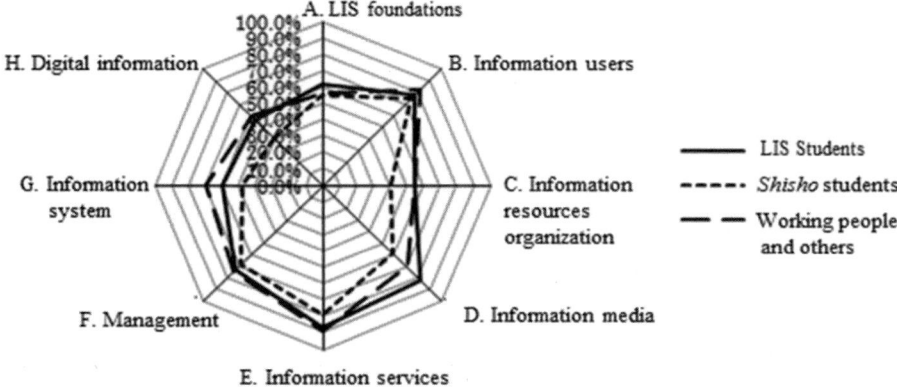

Fig. 4.6 Subject field score distribution of three groups of examinees

4.11   Conclusion

I believe that LIS education should be required even if the library has evolved into a combination of information networks and content management systems without physical settings or physical materials. The education program should be based upon the model of a knowledge information system, which LIS has long fostered, even if the media were digital, printed, or a combination of the two. It is important that the idea and concept of LIS continue to play a role in our society.

Japanese society has recently begun to reconsider the importance of libraries and archival resources. For example, Dr. Makoto Nagao was appointed as NDL Director from 2007 to 2012 based on the bureaucratic personnel rule. This position had been at a ministerial level in the national administrative hierarchy prior to the recent amendment of the NDL Act. For more than 50 years prior, former directors-general of the House of Representatives and House of Councilors had been appointed as NDL directors interchangeably. Dr. Nagao has been the director of the Kyoto University Library. He was also a pioneering researcher of knowledge engineering who developed procedures for analyzing Japanese textual information. Although he is not a librarian, Dr. Nagao has been very understanding of LIS. After resigning, his successor was Mr. Noritada Otaki, who was previously personnel at the Library. This shows that the custom of appointing the director has changed.

Another example is the Public Document Management Act that was enacted in 2009. The National Archives of Japan (NAJ) was established in 1971. However, because precise legal procedures for taking over public documents from administrative units to the NAJ had not been established, it was unable to adequately collect public documents and manage them as national archives collection. When this act was enforced, it empowered the NAJ to achieve the goals of a modern national archive and thus enhanced the nation's administrative and bureaucratic transparency.

The third example, which is currently in progress, is writing the legislation of *gakko-shisho*. As discussed above, there are many *gakko-shisho* who are either regular or irregular employees but the title remains informal because no laws regulate their presence. The Liberal Democratic Party (LDP) was preparing the legislation while they were out of power in 2012 in response to the claims of interested organizations. Since they returned as the ruling party, it has been reported that they actually restarted the legislation process.

These examples show how some changes in the library and LIS environments and LIS scenes are slightly shifting to more political ones in Japan, while they used to be considered apolitical. This may be challenging and ultimately dangerous with little room for even the slightest mistake.

Our society has been steeped in a standard narrative according to which important information was exchanged among highly literate and educated administrative elites. Literacy was high but inequitably distributed. Libraries and archives are the social devices to fill the general literacy and information literacy gaps among people and to achieve social welfare overall. In the twenty-first century, we have begun to appreciate the idea of modern librarianship and LIS education.

These experiences in Japan provide some useful suggestions to international forms of LIS education in Asian and Pacific countries. There are historical reasons that neither the American nor British models of LIS education were pervasive in these countries. Thus, we have to reevaluate and examine alternative possibilities. We must also be skeptical about the dominance of digital technology and culture. National economic development worldwide has been broadly realized through the educational development of populations. Of course, information literacy is needed in every country but it should be based on general literacy. We cannot have information literacy without literacy. We must not forget to aim for a balanced development of technology and culture.

References

Dore, R. (1965). Appendix I school attendance at the end of the Tokugawa period. In *Education in Tokugwa Japan*. University of California Press.

Korotayev, A., Malkov, A., & Khaltourina, D. (2006). *Introduction to social macrodynamics: Compact macromodels of the world system growth* JSPS. Editorial URSS.

Miwa, M., Ueda, S., Nemoto, A., Oda, M., Nagata, H., & Horikawa, T. (2006). Final results of the LIPER project in Japan. World Library and Information Congress: 72nd IFLA General Conference and Council: "Libraries: Dynamic Engines for the Knowledge and Information Society," Seoul, Korea. Retrieved, from http://www.ifla.org/IV/ifla72/papers/107-Miwa-en.pdf.

Nemoto, A. (2009). Galápagos or an isolated model of LIS educational development? A consideration of Japanese LIS education in the international setting. *Asia-Pacific Conference on Library & Information Education and Practice (A-LIEP 2009)*, 6–8 March, 2009, University of Tsukuba, Japan.

Rubinger, R. (2007). *Popular literacy in early modern Japan*. University of Hawai'i Press.

Weber, M. (1990, c1972). Wirtschaft und Gesellschaft: Grundriss der verstehenden Soziologie, Mohr.

Chapter 5
Qualitatively Maintaining Library and Information Science Education in China

Changqing Li

5.1 Introduction

The Chinese political movement known as the Great Proletarian Cultural Revolution (GPCR) ended in 1976 after 10 continuous years. From 1978 onward, China implemented its Reform and Opening-up Policy, prompting rapid development of the nation's universities, their scale to increase exponentially, and student recruitment to expand in unprecedented numbers. These developments marked the transition of Chinese higher education from an opportunity available solely to the elite, to one accessible by the masses.

In parallel with these changes, the field of library and information science (LIS) in China developed. The emergence of new information technologies, particularly the Internet, and societal changes from the early 1990s onward, marked the beginning of a difficult phase for LIS education globally: individuals began to demand a higher standard of LIS education, and China was no exception. Since then, a multitude of government-led initiatives have been enacted to qualitatively reform LIS education; simultaneously, the official recognition of librarians' vocational qualifications has been observed.

Therefore, this paper first discusses major developments that have occurred in Chinese higher education since 1978, how its LIS education has developed, and lastly provides a summary of LIS education's main forms in the country. Based on that, the current state of qualitative LIS education reforms is verified, and trends in the vocational qualifications of librarians are considered. By addressing these topics, the atmosphere necessary for the development of LIS education in China, initiatives required to qualitatively reform the field, and the ideal vocational qualifications of librarians are identified. Some consideration will also be given to the expected direction of LIS education's development in the future.

Changqing Li (✉)
Peking University, Beijing, China
e-mail: licq@pku.edu.cn

M. Miwa and S. Miyahara (eds.), *Quality Assurance in LIS Education:*
An International and Comparative Study, DOI 10.1007/978-1-4614-6495-2_5,
© Springer Science+Business Media New York 2015

5.2 The Development of Higher Education and LIS Education in China Since 1978

5.2.1 The Development of Higher Education

During the GPCR, higher education in China was dealt a blow so severe that most universities were unable to provide a comprehensive education, or conduct research. Moreover, the nationwide university entrance exam was abolished. However, following the end of GPCR in December 1977 the university entrance exam was reinstated after a nearly 10-year hiatus. That year, about 5.7 million people took the entrance exam; among them, 270,000 people passed and subsequently entered a university (Xinhua News, 2009). In the following year, the graduate school entrance exam was reinstated and 63,000 people attended, of whom 10,708 succeeded, although just 18 passed the doctoral exam (China Education Yearbook, 1984, p. 964). This signified that postgraduate education had also returned to a state of normalcy. Incidentally, in 1977 only 404 universities existed in the entire country (China Education Yearbook, 1984, p. 965).

University education recovered relatively smoothly in subsequent years, and the number of students increased steadily each year; by 1997, for the first time ever, more than one million candidates, including junior-college students, successfully began their university studies. In 1999 higher education in China reached a second major turning point; that year, China's National Development and Reform Commission and the Ministry of Education issued an urgent communique to all universities demanding increased student enrollment. As a result, enrollment increased enormously, and the number of students accepted to attend 4-year programs rose to 1,118,400. Also, of the approximately 320,000 individuals who took the graduate school entrance exam, 92,200 people passed. In a period of 20 years the number of universities, including junior colleges, climbed to 1,846—a more than fourfold increase since 1978 (China Education Yearbook, 1984, p. 94).

The enormous increase in student enrollment at universities had major repercussions for Chinese society, allowing more young people to realize their dream of higher education. At the same time, however, it placed an enormous burden on the higher educational system, and research and debate continues even today concerning whether this expansion resulted in a decrease in the quality of Chinese higher education. Due to an increased number of universities and student enrollment it is clear that higher education in China has transitioned from an opportunity available to only the elite, to something obtainable by the masses. By 2011, the number of universities and junior colleges rose to 2,409 institutions and 6.81 million students—a sixfold increase from 10 years earlier. In 2013, there were 560,000 postgraduate enrollments, 65,000 of which entered a doctoral program, representing an enrollment rate of 56 times greater than 30 years prior (Ministry of Education, 2013, July 30).

5.2.2 The Development of LIS Education

The development of LIS education in China followed a path similar to the growth of higher education. Throughout the 6-year span of the GPCR, library science education (LSE), like other university programs, had been almost completely suspended and was almost abolished. Not only were there no students enrolled in LIS education courses, but its teachers were dispersed, materials scattered or lost, and specialized research halted. In 1970, many specialist programs and their enrollment resumed, although library science was not one of them. Peking University and Wuhan University resumed enrollment of library science students in 1972, albeit only for a 2-year course. One year after the Gang of Four's removal from power in 1975, Peking University and Wuhan University lengthened the duration of their library science programs to 4 and 3 years respectively. Later, in 1978, Wuhan University also extended the duration of its library science program to 4 years, marking the return of LSE in China to its prior pre-revolution status.

It was from 1978 onwards that LSE in China expanded its scale and developed considerably. Two factors contributed to this. The first was that after 1978 China, on average, opened a new public library every 3.7 days; (Editorial Board of Contemporary Chinese Library Cause, 1995) consequently, individuals with library science training were in high demand. Secondly, during the GPCR, there were many librarians who lacked specialized instruction, resulting in a significant growth in on-the-job training and adult education.

1978 also marked the year that Wuhan University's Department of Library Science was established, and a major in science and technology information science was introduced. Accordingly, other universities such as Peking University and Nanjing University sequentially established information science majors; as a result, library science departments at many other universities in the early 1980s were renamed to either "library and information science," or library "information science."

In the span of just 10 years between 1978 and 1987 the number of LIS education departments throughout China had risen from 2 to 52, and the number of students attending a 4-year library science program had increased from approximately 200 students in 1977 to 6,300 students in 1987 (Zheng, Li, & Wang, 2001, p. 65). During the same period, LSE in graduate schools reemerged and expanded. In 1966, prior to the GPCR, Peking University's Department of Library Science had two postgraduate students enrolled in its master's course before being completely eliminated. From 1978 to 1979 at Wuhan University and Peking University respectively, postgraduate students were enrolled in a revised 3-year degree program. In 1982, both universities began enrolling postgraduate students in a program comprising 2 years of coursework; two additional years were allotted for the completion and submission of a thesis, after which students would receive a master's degree. By 1988 six institutions in China offered a master's degree in library science, while seven offered a master's degree in information science. Between 1978 and 1987,

147 postgraduate students completed LIS education programs, while 221 students had attended some form of graduate studies (Zhang, Xun, & Shen, 1989, p. 617). In 1990, with approval from China's State Council doctoral programs in LIS were established at Peking University and Wuhan University respectively. Subsequently, Wuhan University, the Document and Information Centre of the Chinese Academy of Science (DICCAS) and Nanjing University established doctoral programs in library science, while Peking University established doctoral programs in information science. According to statistics, in 1995 58 universities throughout China had established LIS programs, with a total enrollment of more than 4,000 students. Of those universities 53, 11, and 3 offered undergraduate, master's, and doctoral programs respectively (Huang, 1999, p. 9). By 1998, 1,460 and 43 postgraduate students were enrolled in LIS master's and doctoral programs respectively (Peng, 2000, p. 33).

In September 1992, Peking University was the first to adopt the moniker "information management" for its information science department, which had major repercussions throughout the country. Following this change, by March 1995, 26 of the 52 facilities across China with LIS departments had followed suit, or adopted similar titles such as information resources management, information technology decision sciences, document information management, and information industries (Dong, 1996). In response to this, some changes were also made to their curricula and offered courses. The impetus for these name changes included the three following factors. First, the Chinese word for intelligence was slowly replaced by the term "information" in popular usage. Secondly, students searching for employment in either the public or private sector would have an undeniably easier experience locating work with a more generically named degree such as "information management." Thirdly, a series of library science school closures in the USA had a major impact on the field in China. To remain relevant and maintain student enrollment, these facilities realized that it would be easier to survive as LIS education providers if they changed their names to a more contemporary form.

5.3   The Main Types of LIS Education Offered in China

The hub of LIS education in China today is the Department of Information Management at Peking and the Wuhan University's School of Information Management; there are approximately 60 information management departments at universities and junior colleges across the country, in addition to various information management and information systems programs and departments. Peking University and Wuhan University are pioneers in this field, predating most other information management programs and departments that were not established until after 1978. The students in this field include those enrolled in university departments, graduate schools, and distance-learning programs. Universities generally provide the four LIS education programs discussed below.

5.3.1 Standard 4-Year Undergraduate Program

Standard 4-year LIS education undergraduate programs still occupy an important position in China. In July 1998, China's Ministry of Education (MOE) distributed a new catalog of undergraduate courses and majors that included a short description of each. The five information science majors previously known as Science and Technology Information, Economic Information Management, Management Information Systems, and Forestry Information Management were merged under the umbrella of Information Management and Information Systems. By 2007, 366 facilities across the country had established Information Management and Information Systems programs that were usually located within a computer or computer science department (Qiu, Ni, & Meng, 2008, p. 11). In recent years, the number of schools that retained majors under the title of Library Science declined from 30 to 26 (Wang et al., 2010). Peking University and other schools offer students an undergraduate course of study named Information Management and Information Systems, although students are offered a choice between a specialization in either "Library Science" or "Information Management and Information Systems" in their junior year. According to a survey conducted between 2003 and 2006 at 23 different facilities offering undergraduate degrees in this field, including Peking, Wuhan University, and Sun Yat-sen University, there was an overall increase in undergraduate enrollment to LIS programs, reaching a total of 3,800 students (Chen et al., 2008, p. 8). Also, based on the results of a different survey of 23 universities in 2009, the total number of undergraduate students enrolled in an LIS program was 1,125 students indicating that, on average, 49 undergraduate students were recruited per university with 654, 361, and 100 students enrolled in liberal arts, engineering science, and departments not classified under either liberal arts or engineering science respectively (Chen et al., 2009, p. 5).

5.3.2 Postgraduate Programs

As described earlier, in numerical terms a large percentage of individuals enrolled in Chinese LIS education programs are full-time undergraduate students. In recent years, however, significant developments have been made in postgraduate education, and the focus of LIS education is gradually shifting from undergraduate to postgraduate studies. LIS postgraduate education in China can be divided into three levels: master's coursework, doctoral coursework, and graduate classes. In 2007, 67 and 38 universities offered an information science and library science master's degree respectively. Additionally, 16 facilities offered doctoral degrees: eight in library science and the remaining half in information science (Qiu, Ni, & Meng, 2008, p. 11). Due to a sharp increase in the number of graduate schools offering LIS education, the student enrollment in these programs also grew significantly between 2002 and 2006, the 30 facilities offering a master's degree in library science reported an enrollment of 1,291 postgraduate students; this number grew from 178 students in 2002 to 338 students in 2006, signifying a cumulative increase of 89.9 % and an average increase of 22.5 % per year (Xiao, Si, & Huang, 2008). Currently, Peking University enrolls approximately 50 students to its 4-year undergraduate program each year; comparatively, it enrolls on average 35 and 12 postgraduate students to its master's and doctoral programs respectively.

With regard to the length of the postgraduate programs, master's degrees at Peking, Wuhan University, and Nanjing University take 2 years to complete, while other universities offer 3-year degrees. The completion of a doctoral degree usually requires 3–4 years, although in 2006 Peking University standardized upon a 4-year program.

In accordance with a series of recent reforms to postgraduate studies, China's MOE has reduced the number of facilities offering general academic degrees in favor of specialized degrees. The formal introduction of a master's degree in Library and Information Science (MLIS) in 2010 is one example of this, whose enrollment comprises mainly adult learners and university students seeking specialized human resource training with a focus on high-level workplace skills rather than academic research. As of June 2010, 18 universities, including Peking University and Wuhan University have established MLIS programs (Ke, Zhang, & He, 2012, p. 92).

In addition, a program exists for postgraduate students who are currently employed but seeking to further their education. This is ordinarily a 2-year program comprising ten subjects such as foreign language studies, philosophy, and individual courses from the LIS master's program. After successfully passing a final exam, participants receive a certificate affirming their completion of a postgraduate program. Unlike other postgraduate students, these individuals are not designated an academic advisor nor required to write a graduation thesis during their time of enrollment. After completing their coursework, however, students are expected to begin writing their master's thesis while continuing their day-to-day jobs and submit it to the university for review.

In 2003, China's Ministry of Human Resources and the Nationwide Postdoctoral Administration Committee designated 434 educational institutions across the country to accept postdoctoral students; subsequently Peking University, Wuhan University, and Renmin University, China established research positions for LIS, Public Record Science, and Document Information Management (Chen et al., 2008, p. 7). In 2007, a postdoctoral program was also initiated at Nanjing University's Department of Information Management. By sponsoring postdoctoral research, China is able to maximize the talents of its workforce.

5.3.3   Distance and Online Education

Since the 1980s distance learning[1] has gained rapid adoption in higher education, and presently employed students majoring in LIS account for a significant portion of those enrollees. Distance education in China takes either a specialized or general form, and requires 3 years to complete in both cases. Applicants to a specialized program must at minimum possess a high-school diploma; in contrast, applicants to the general program should have graduated from the specialized program or hold equal or higher academic credentials. Since distance learners do not attend a traditional classroom, each student attends immersion programs and is assigned an advisor to meet with regularly.

Peking University and Wuhan University began concurrently offering undergraduate-level distance-learning programs in 1985. Incomplete statistics collected between 1980 and 1990 reveal that approximately 6,000 students participated in distance education (Editorial Board of China Library Yearbook, 1997, p. 359). Even today, institutions such as Peking University and Wuhan University target adult members of the workforce for enrollment in both junior college and undergraduate distance-education programs.

Also in 1985, following approval from the MOE, the Open University of China established a library science department and enrolled more than 20,000 students in its first year (Ke, 1986, p. 69). In recent periods, however, portions of the department have ceased to accept new applicants.

Upon entering the twenty-first century, the maturation of the Internet and other information and distance-learning technologies prompted many Chinese universities to establish online classes, whose enrollees include adult students or younger students who failed their university entrance exams. After completing the necessary coursework and submitting a graduation thesis, students are awarded a Bachelor's degree. According to recent statistics, Peking University's Department of Information Management enrolls approximately 1,000 online students each year.

5.3.4   Apprenticeships

Generally, apprentices are selected by university instructors and library employees with the intention to increase their research skills and work capabilities. The duration of an apprenticeship is most commonly from 6 months to 2 years, and entails either specialized research or the acquisition of knowledge vital to the individual's duties.

---

[1] In distance education (函授教育), (A) working students who graduated high school or equivalent education take national standard examination for distance education, (B) combines independent self-regulated learning and intensive schooling for face-to-face classes using standardized textbooks, (C) divided into short program (three years) and regular program (four years); only a limited number of excellent students are eligible to receive bachelor's degree, (D) completion of the program is certified by the country.

5.4 Qualitatively Reforming Chinese LIS Education

Since the 1990s, Chinese higher education has developed rapidly, and the massive increase in undergraduate and postgraduate student enrollments each year is on an unsurpassed level when compared to other nations. This condition is mirrored in the country's LIS education, where increased enrollment and relative size requires perpetual adjustments to ensure the provision of a quality education. Below, the fundamental measures necessary to qualitatively maintain LIS education as both a traditional and a contemporary specialization in China are introduced.

5.4.1 Traditional Library Cataloging as an Academic Discipline and the Establishment of LIS

China has practiced and refined the science of library cataloging for more than 2,000 years. Its practice and theory is a valued discipline in Chinese academia, and it is an important tool for scholars and students alike to familiarize themselves with as an area of study. In recent years, traditional Chinese library sciences have experienced a resurgence, and its effect upon LIS education is evident. A commitment to precision and perfection is fundamental to library cataloging, and these are also qualities that dedicated scholars are required to possess in the spirit of searching for and single-mindedly studying the *shintai*, or ultimate truth.

More than 90 years have passed since China formally opened its first library science school in 1920. While its operation has been occasionally disrupted due to wars or political upheavals such as the GPCR, LSE and subsequently LIS education continued to evolve and remain accessible to students throughout the years. China's historical commitment to LSE and LIS education, and the contributions that its experts have made to Chinese society are widely known. As a field of learning with a long history of quality and excellence the future of LIS education has already been secured to a certain extent.

5.4.2 Implementing a Major in Library Cataloging and Establishing LIS as an Esteemed Subject

In July 1993, the MOE and the Bureau of Higher Education published a guide to the majors and courses offered by the country's institutions of higher learning. The guide provided an overview of majors available, their names, categories, and subcategories and also set standards for work and employee training. More pertinent to this chapter, perhaps, are its requirements for the implementation and coordination of majors; this addresses issues such as training specialized staff members, degree requirements, new student recruitment, job placement for recent graduates, educational statistics, and human resource forecasting. Essentially, it is an official state document containing directives for the macro-management of higher education (Higher Education Department of Ministry of Education of the People's Republic of China, 1993). A revised edition of the guide updated the names of available library science and information science majors in accordance with the specifications of relevant departments. This edition also marked the formal name change of information science to information management. Prior to this revision, library and information sciences were categorized as primary subjects belonging to scientific fields. After the revision, however, library, information, public record, and book publication science were collectively categorized as The Science of Books, Information, and Public Records and were no longer consider a scientific, but historic academic discipline.

Also in 1993, the Chinese National Council Degree Committee published a guide to postgraduate LIS education that no longer categorized library and information sciences as subcategories of history and science, but as primary subjects. In 1997, a revised edition of an official document concerning master's and doctoral degree program requirements and majors established Library, Information, and Public Records Management (LIPRM) as a primary major under the umbrella of management studies. Prior to that in 1994, for the first time ever, the National Social Science Fund Assistance Project allocated greater resources to library, information, and public records sciences; this indicated that LIS research remained a high priority for national research.

By issuing guidelines for undergraduate and graduate studies, LIS acquired a status equivalent to other subjects such as philosophy, law, and mathematics, which have all contributed to the development of LIS in China. The establishment of LIS as an esteemed and well-respected subject has played a major role in ensuring that the quality of LIS education will remain consistent for years to come.

5.4.3 Stipulating Guidelines for the Initiation of Primary Doctoral Degrees and Their Designation as Subjects of National Importance

In 2000, the Chinese National Council Degree Committee deemed Peking and Wuhan University's LIPRM program an official primary doctoral degree; later, in 2006, Nanjing University and Renmin University of China's LIPRM programs acquired the same status. Also that year, the number of institutions qualified to offer a secondary master's degree in the field more than doubled to 127, compared to 57 six years earlier. Of these 127 institutions, the number offering master's degrees in library science and information science increased by 20 and 40 respectively (Chen et al., 2008). Universities granted approval to initiate primary doctoral degrees are capable of creating new programs independently to meet fluctuating scholarly and societal demands. Subsequently, Peking University and Wuhan University have both established postgraduate programs in publishing studies and information resource management. This hierarchy has allowed China to maintain the structure of its LIPRM programs and departments in a more rational manner.

In 2002 and 2007, preparations to select new subjects of national importance occurred; as a result, Peking and Wuhan University's library science programs were chosen both years, in addition to Wuhan University's information science program. In 2007, Nanjing University's information science program was also deemed nationally important. The state made these selections by considering factors such as the number of teachers employed within each program, student enrollment, and research output. A program's status as a subject of national importance is an important indicator of a university's overall quality of education.

5.4.4   An Introduction to China's Approach to Program Evaluation

Three phases are used to evaluate a university's provisioning of undergraduate and postgraduate education in China. The first phase comprises an evaluation of a school's undergraduate program performed by the MOE; the second phase entails an evaluation conducted by a subdivision of the MOE, and focuses solely on an institution's postgraduate studies. In the third phase, other universities and private-sector organizations rank each postgraduate and undergraduate program accordingly.

On a number of occasions throughout the 1980s, the MOE evaluated universities' provisioning of undergraduate education. In August 2004, however, the MOE established the Higher Education Evaluation Center of The Ministry of Education, a dedicated subdivision that introduced a system of quinquennial evaluations. This was significant since it demonstrated a standardized, scientific, systemized, and specialized effort to evaluate Chinese higher education. Consequently undergraduate library and information science programs were subject to these evaluations as well.

The Higher Education Evaluation Center of the Ministry of Education began its evaluation of postgraduate education in 2002, and by 2012 it had carried out three rounds of assessments. Each primary program was evaluated upon its teachers and educational environment, quality of research produced, employee training, and the program's prestige. In the third round of evaluations conducted in 2012, Wuhan University's LIS program was given top ranking.

The evaluations and subsequent rankings of individual schools and their available programs compiled by other universities and private-sector organizations have a decidedly strong influence on students' likelihood to apply for a specific university. However, private-sector evaluations have a limited effect on the universities' decision making or the direction of their programs. While there is no consensus on their accuracy, and while each system may have its unique problems, the evaluations are still believed to play a positive role in enhancing the overall quality of education provided by specialized programs.

5.4.5 Criteria for the Establishment and Selection of Ideal LIS Education Courses

In 2002, the MOE's Guidance Committee for Library Science Learning and Education in Institutions of Higher Learning was established to provide input on the standardization of LSE throughout the country. Based on this guidance, many institutions with LIS education programs determine the courses they intend to offer and differentiate between required and elective classes.

China widened its educational reforms in 2003, and to generate enthusiasm, the MOE launched a national program showcasing departments with the most exemplary teaching staff, management, content, methods, and materials. It was hoped that the departments and individuals showcased by this initiative would serve as role models for other educators. Between the program's inception and 2009 the LIPRM departments of nine different universities were featured, including those at Peking University, Wuhan University, and Renmin University of China. Additionally, Wuhan University's teaching team teaching core curriculums in library and information science was showcased in 2007 due to the excellent track record of its LIS undergraduate program.

The aforementioned systems and measures are intended to increase the overall quality of higher education and assist in the maintenance, stabilization, and improvement of LIS education in China.

5.5 Professional Qualifications of Chinese Librarians

5.5.1 Job Promotions Among Chinese Librarians

A common hierarchy of librarians exists in China with both generalized and specialized staff members; specialized staff members can be classified into the following five categories: librarians, assistant librarians, administrators, assistant researchers, and researchers. In this system, staff members are promoted to specialized duties based on a cumulative review of their academic credentials, work abilities, and work achievements. For example, a junior college graduate may qualify for an administrative position after completing a 1-year apprenticeship; similarly, university graduates or those with an equivalent academic background can become assistant librarians following a year-long apprenticeship. After obtaining an additional 5 years of experience, assistant librarians are eligible for promotion to administrators, and then those eligible individuals will become assistant researchers. Finally, after another 5 years of service, an assistant researcher is qualified to become a full researcher. These standards for promotion have been in effect by decree since 1981, and after more than 30 years, it remains a deeply rooted process applicable to any librarian irrespective of his or her qualifications. This has sparked debate in recent years concerning the importance of vocational qualifications among librarians, and whether the current criteria for promotion should be reevaluated.

5.5.2   Trends in Librarian Qualifications

Since 2001, the Chinese Society for Library Science (CSLS) has conducted research, consulted experts, and gathered data from countries such as the UK, the USA, Japan, and Australia concerning the professional qualifications of librarians. Based on this information, it carried out a variety of investigations from both an academic and practical point of view. In August 2002, the CSLS submitted a report to the Ministry of Culture (MOC) identifying important vocational qualifications for the country's librarians. In February of the following year, a summary report was sent to the Bureau of Human Resources and Bureau of Society, Culture, and Libraries; in it, the CSLS provided a comprehensive overview of vocational requirements for librarians in key industrialized nations. During this same period, the society gathered experts for a symposium in Beijing to discuss the implementation of a national qualification standard.

In March 2003, following a request from the MOC's Vocational Qualifications Appraisal and Guidance Center, the CSLS gathered experts from a variety of fields to create a draft proposal of national standards for three groups of librarians specializing in reading materials, classical documents, and document restoration. In November of the same year, the proposal was submitted to the Ministry of Labour and Social Security and to the MOC who formally approved the standards in July 2004. The finalization of these standards laid the groundwork for the CSLS to begin compiling teaching materials for the purpose of training and familiarizing library staff and specialists in reading materials, classical documents, and document restoration with the newly implemented vocational requirements. This task was completed in 2005.

A year later in May 2004, Sun Beixin, the former assistant director of the National Library of China sponsored additional research into the establishment and implementation of a vocational qualifications system for Chinese librarians. This research was in turn a key component of the National Social Sciences Fund Assistant Project, which contributed to the fundamental philosophical and regulatory elements that would eventually become a vocational qualification system for the country's librarians.

5.5.3   Implementation and Enforcement of Professional Standards for Reading Materials Specialists

The official trial implementation of the national professional standard for reading materials specialists began on July 27, 2004. Reading materials specialists' primary responsibilities include collecting, arranging, distributing, managing, and developing document information in addition to other related duties. Reading materials specialists are divided into five levels in accordance with the national vocational qualification guidelines. The professional skills required for this position include acute observation, sound judgment, clear communication skills, and linguistic competence. With regard to their educational backgrounds, employees must be high school graduates or have completed an equivalent level of education.

Employee levels are determined by the individual's training. For example, a reading materials specialist is a level 5 employee if he or she has spent a minimum of 200 h in the classroom; similarly level 4, 3, 2, and 1 employees must accumulate a minimum of 240, 260, 240, and 220 h respectively. There are also regulations in place for staff trainers; trainers who oversee reading materials specialists at levels 5, 4, and 3 must possess at minimum a level 2 qualification or hold a higher position, such as a library staff member. Trainers of level 1 and 2 reading materials specialists must have a level 1 qualification and 5 or more years work experience after obtaining their level 1 qualification; alternatively, they must occupy a high-level position such as a researcher.

The following fundamental regulations were also established before an individual can apply for higher qualifications:

If applying for a level 5 qualification, the applicant must belong to one of the following three groups:

1. Individuals who have graduated with a relevant major from a secondary vocational school and have worked continuously for at least 1 year in an LIS-related profession
2. Individuals with at least 2 years of continuous LIS-related work experience
3. Individuals who have received level 5 training, completed the prescribed number of classroom hours, and acquired a certificate of qualification

When applying for a level 4 qualification, the applicant must belong to one of the following four groups:

1. Individuals who, after acquiring their level 5 qualification, have worked continuously in an LIS-related profession for 3 years or more
2. Individuals who have graduated with a relevant major from a 3-year junior college program and have worked continuously in an LIS-related profession for a year or more
3. Individuals who have worked continuously in an LIS-related profession for 8 years or more
4. Individuals who, after acquiring their level 5 qualification, have worked continuously in an LIS-related profession for 2 or more years; received level 4 training; completed the prescribed number of classroom hours, and acquired a certificate of qualification

When applying for a level 3 qualification the applicant must belong to one of the following three groups:

1. Individuals who, after acquiring their level 4 qualification, worked continuously in an LIS-related profession for 3 years or more
2. Individuals who have graduated with a relevant major from a 4-year university program and have worked continuously in an LIS-related profession for a year or more

3. Individuals who, after acquiring the level 4 qualification, worked continuously in an LIS-related profession for 2 years or more; received level 3 training; completed the prescribed number of classroom hours, and have acquired a certificate of qualification

When applying for a level 2 qualification, the applicant must belong to one of the following three groups of people:

4. Individuals who, after having acquired the level 3 qualification, worked continuously in an LIS-related profession for 5 years or more
5. Individuals who have completed a relevant major or postgraduate master's course and have worked continuously in an LIS-related profession for 2 years or more
6. Individuals who, after acquiring the level 3 qualification, worked continuously in an LIS-related profession for 3 years or more; received level 2 training; completed the prescribed number of classroom hours, and acquired a certificate of qualification

When applying for a level 1 qualification, the applicant must belong to one of the following three groups of people:

1. Individuals who have an educational background equivalent or superior to a relevant major from a 3-year junior college program and who, after having acquired the level 2 qualification, worked continuously in an LIS-related profession for 5 years or more
2. Individuals who have earned a relevant postgraduate doctoral degree and have worked continuously in an LIS-related profession for a year or more
3. Individuals who, after acquiring the level 2 qualification, worked continuously in an LIS-related profession for 3 years or more; received level 1 training; completed the prescribed number of classroom hours, and have acquired a certificate of qualification

The above criteria describe the training that must be completed by reading materials specialists and the applicants' required qualifications. Efforts are being made for this system to be adopted by libraries throughout China.

5.5.4   Problems Facing the Vocational Qualifications System

Despite the fact that the Chinese Government has officially endorsed the standards discussed in this chapter, a number of major problems concerning the vocational qualifications system remain.

The first issue is that the national standards for reading materials specialists are identical to the education and training already held by those working in libraries; therefore, the qualifications are not necessarily meaningful. In this regard, the system does not resolve the question of whether a person seeking work at a library actually requires a specialized qualification.

Secondly, another major problem concerns the conflict between the contemporary qualifications system and the earlier system of promotion that has existed for three decades. There are a number of opinions regarding what should occur following the implementation of a qualifications system. One side, for example, argues that the new system can operate in tandem with the old: veterans are still governed by the old system, while newcomers adhere to the contemporary system. Regardless, this problem requires ample time and creative thinking to be properly resolved.

The third problem concerns how to best integrate the vocational qualification system with the previously established theories of LIS education. Should the training of reading materials specialists be performed by universities' LIS departments, the MOC's Vocational Qualifications Appraisal and Guidance Center, or by the vocational schools that exist in abundance throughout China? This question remains unanswered. Nevertheless, the training that reading materials specialists undergo must entail more than the provision of vocational education and training. Indeed, training and retraining librarians is a task with an enormous economic bounty involved, and the providers of these services have a great financial stake in its outcome. China's obsession with the acquisition of various academic qualifications has been commercialized to its fullest extent in recent years, and the market is oversaturated. If the training and education of reading materials specialists is allowed to deviate drastically from its intended purpose, it will serve no benefit to the establishment of a vocational qualifications system for librarians.

Fourthly, China's library system comprises public, university, and specialist libraries. Each system is overseen by a different governmental division: public libraries are the domain of the MOC, university libraries of the MOE, and specialist libraries of the Ministry of Science and Technology. The national professional standards for reading materials specialists are selected and approved by the MOC and the Ministry of Labour and Social Security, while the MOE and the Ministry of Science and Technology do not participate in this process. Therefore, it remains unclear whether the national standards for reading materials specialists will be adopted throughout the country or merely at public libraries.

Lastly, although a number of topics have been addressed in this text, important issues such as the creation of a nationwide exam to evaluate librarians' vocational qualifications, the effective period of the qualifications, and their recognition or dismissal by the library directors, going forward, will be necessary to be critically examined.

5.6   Conclusion

Since 1978, LIS education in China has grown enormously to become an indispensable component of higher education and an active contributor to the nation at large. However, as information technology and society rapidly evolve, LIS education is encountering new challenges on an unprecedented scale. Reforms to information science education, revisions of educational content, the restructuring of curriculums, and the establishment of a vocational qualifications system for librarians are collective threats that could potentially destabilize LIS education. Although the Chinese government has been actively attempting to address these problems, a number of questions remain concerning LIS education's ability to simultaneously evolve and ensure excellence. To overcome these hurdles, LIS researchers and those involved in LIS education must collaborate to continue moving forward.

References

Chen, C., et al. (2008, September). The development of library and information science education in China from 1978 to 2008. *Documentation, Information and Knowledge, 9*(125). (Source in Chinese: 陈传夫,吴钢,唐琼, 孙凯,于媛. 改革开放三十年我国图书情报学教育的发展[J]. 图书情报知识, 2008, 5:5–14.)

Chen, C., Wu, G., Sheng, Z., Ding, N., & Zhang, F. (2009). The development and prospects of library and information science education in Chinese main land from 1949 to 2009. *Library Journal, 8*(28). (Source in Chinese: 陈传夫,吴钢,盛钊,丁宁,张法. 新中国图书情报学教育历程与展望[J]. 图书馆杂志, 2009, 8:3–11.)

China Educational Yearbook. (1984). *China education yearbook 1949–1981*. The Editorial of China Education Yearbook. Beijing, Encyclopedia of China Publishing House, p. 964 (Source in Chinese:中国教育年鉴编辑部. 中国教育年鉴1949–1981[M]. 北京：中国大百科全书出版社, 1984, p. 964.)

Dong, X. (1996). Pattern change and problems of Chinese library and information science education. *Journal of Library Science in China, 22*(101), 30. (Source in Chinese: 董小英. 我国图书馆学情报学教育的转型及其问题[J]. 中国图书馆学报, 1996, 1:30.)

Editorial Board of Contemporary Chinese Library Cause. (1995). *Contemporary Chinese library cause* (Contemporary Chinese books) (pp. 56–57). Beijing: Contemporary China Publishing House. (Source in Chinese:当代中国的图书馆事业编辑委员会. 当代中国的图书馆事业[M]. 北京：当代中国出版社, 1995, pp. 56–57.)

Editorial Board of China Library Yearbook. (1997). *China library yearbook* (p. 359). Beijing: National Library of China Publishing House. (Source in Chinese: 中国图书馆年鉴编辑委员会.中国图书馆年鉴[M].北京：北京图书馆馆出版社, 1997, p. 359.)

Higher Education Department of Ministry of Education of the People's Republic of China. (1993). *Undergraduate professional directory and professional colleges and universities*. Beijing: Higher Education Press (HEP). (Source in Chinese: 国家教育委员会高等教育司编.普通高等学校专业目录和专业简介[M].北京：高等教育出版社, 1993.)

Huang, Z. (1999). The Chinese librarianship of 20 years of reforming and opening to the World. *Library*, no. 2, 9. (黄宗忠,黄力. 改革开放20年的中国图书馆事业[J]. 图书馆, no. 2, 1999, p. 9.)

Ke, H. (1986). Overview of library science professional of China Central Television University. *Documentation, Information and Knowledge*, no. 2, 69. (Source in Chinese: 柯寒. 中央电大图书馆学专业概述[J]. 图书情报知识, no. 2, 1986, p. 69.)

Ke, P., Zhang, W., & He, Y. (2012). Reflections on some issues concerning the degree education of library and information science in China. *Information and Documentation Service,* no. 6, 92. (Source in Chinese: 柯平, 张文亮, 何颖芳. 对我国图书情报专业学位教育若干问题的思考[J]. 情报资料工作, no. 6, 2012, p. 92.)

Ministry of Education (2013, July 30). Ministry of Education of the People's Republic of China. (Source in Chinese: 中华人民共和国教育部). Retrieved, from http://www.moe.gov.cn/publicfiles/business/htmlfiles/moe/s7382/index.html.

Peng, F. (2000, January). Graduate education in library and information science in China for the new century. *Journal of Library Science in China*, *26*(125), 33. (Source in Chinese: 彭斐章. 迈向21世纪的我国图书馆学情报学研究生教育[J]. 中国图书馆学报, 2000, 01:33.)

Qiu, J., Ni, C., & Meng, Y. (2008, November). Comparative analysis on the status quo of library and information science. *Documentation, Information and Knowledge*, *11*(126), 11. (Source in Chinese: 邱均平, 倪超群, 孟园. 海峡两岸图书情报学教育现状比较分析[J]. 图书情报知识, 2008, 06:11.)

Wang, Z., et al. (2010). Review and outlook of library and information science education on the 30 years (1978–2008) in China. *Library and Information*, (2), p. 27. (王知津,徐芳,潘永超,王秀香,刘念. 我国图书情报学教育三十年 (1978–2008) 回顾与展望[J]. 图书与情报, no. 2, 2010, p. 27.)

Xinhua News. (2009, August 25). In 1977, Restore the university entrance exam (Source in Chinese: 1977年恢复高考, 新华社, 2009年8月25日). Retrieved, from http://gov.hnedu.cn/web/0/200908/25110121250.html

Xiao, X., Si, L., & Huang, R. (2008, May). A survey of education in library science and professional requirement. *Journal of Library Science in China, 34*(175), 11–16. (Source in Chinese: 肖希明, 黄如花, 司莉. 我国图书馆学专业教育与职业需求的调查与分析[J]. 中国图书馆学报, no. 3, 2008, p. 11–16.)

Zhang, B., Xun, C., & Shen, J. (1989). *Decade of Chinese librarianship, 1978–1987* (p. 617). Changsha: Hunan University Press. (Source in Chinese: 张白影, 荀昌荣, 潘继武.中国图书馆事业十年 (1978–1987) [M].长沙:湖南大学出版社, 1989, p. 617.)

Zheng, Z., Li, S., & Wang, H. (2001). *An introduction to library science education in China* (p. 65). Changsha: National University of Defense Technology Press. (Source in Chinese: 郑章飞, 黎盛荣, 王红.中国图书馆学教育概论[M].长沙:国防科技大学出版社, 2001, p. 65.)

Chapter 6
LIS Education and Quality Assurance System in Asia-Pacific: Taiwan

Chihfeng P. Lin

6.1 Introduction of LIS Education Programs in Taiwan (1961–2012)

6.1.1 Development of LIS Education 1960–2010

The development of education of librarianship in Taiwan has been through more than half of a century since 1954 when the first LIS Education program was initiated in National Taiwan Normal University. The formal education of librarianship inherited from Wen-Hua Library School (now Wu-Han University, China). Colleagues from Wen-Hua Library School moved from China to Taiwan and cooperated with local scholars to establish library education after a decade of efforts. National Taiwan Normal University initiated a Librarian Education Program under Department of Social Science. In the next 10 years, from 1961 to 1970, there were five (NTNU, NTU, SHU, FJCU, and TKU) LIS education institutes that were permitted to conduct the education and training of the profession. Follow-up with the development of the profession, after 20 years of work, these schools changed names into Library and Information Science (Studies) in the last 10 years of the twentieth century.

Currently, there are nine universities that provide Library and Information Science (LIS) education in Taiwan, among these institutes, five have undergraduate program which includes Department of Library & Information Science, National Taiwan University (NTU), Department of Library & Information Science, Fu-Jen Catholic University, Department of Information and Library Studies, Tam-Kang University (TKU), Department of Information & Communications, Shih Hsin University (SHU), Department of Library and Information Science, Hsuan-Chuang University (HCU); Eight institutes have graduate programs which are NTU,

C.P. Lin (✉)
Shih Hsin University, Taipei, Taiwan
e-mail: chihfeng@cc.shu.edu.tw

M. Miwa and S. Miyahara (eds.), *Quality Assurance in LIS Education:*
An International and Comparative Study, DOI 10.1007/978-1-4614-6495-2_6,
© Springer Science+Business Media New York 2015

FJU, TKU, SHU, Graduate School of Library, Information, and Archival Studies, National Cheng-Chi University (NCCU), Graduate School of Library & Information Studies, National Chung-Hsin University (NCU), Graduate School of Library and Information Science, National Taiwan Normal University (NTNU), Digital Library and Information Program, College of Engineering and Information, National Jiao-Tong University (NJTU); Three Universities provide Doctoral Programs, NTU, NTNU, and NCCU. Table 6.1 indicates Institute Name, Location, Year Established, College/School to which it belongs, Existing Programs.

6.2   Current Status of LIS Education in Taiwan

One can identify that the development of LIS Education in Taiwan started with undergraduate programs to fulfill needs of society which was undergoing economy and social development in 1960s. After 20 years, the graduate program for Master's Degree of Library Science was initiated by National Taiwan University in 1980. Following the development of the librarianship, the Department of Library Science of NTU changed its name to Library and Information Science in 1989.

Other Library Schools followed and become Library and Information Science (or Studies) in the following 10 years. NCCU's Master's Degree Program added archival studies in addition to traditional LIS Program. NJTU provided a "Digital Librarian Program" specifically on training information technology for digitizing process of libraries to fulfill nation-wide needs of the digitizing era.

In accordance with the blooming of Librarianship in Taiwan since 1980 while the government developed about 300 public libraries in the country plus the increasing number of universities/colleges which were permitted to upgrade from non-degree community colleges or technological colleges. This was the main driven force of the advance studies, thus, Master Degrees Programs that were established by of graduate schools, SHU, FJU, TKU, NCCU, CHU, NTNU, and NJTU in this decade. In the year of 2000, Department of Library and Information Studies of Shih Hsin University changed its name into "Department and Graduate Programs of Information and Communications" to distinguish the emphasis of Communication theory and practices in information services.

Department of Educational Information Science of Tam-Kang University changed its name into Department of Information and Library Science in 2001 to upgrade its curriculum contents of information.

From 1990 to 2010, LIS Schools have positioned themselves into three categories, the first one is traditional LIS education that produce graduates to serve in academic libraries, special libraries; The second one is training school librarians for elementary school, junior and senior high Schools; The third one is more applicable to public libraries. In addition to above, NCCU's added archival studies which injected new categories of the librarianship; SHU's emphasis on communication theories and practices that allows librarianship to broaden the use of media channels for information services and expand the job market beyond libraries. As for

Table 6.1 Institute name, location, year established, college/school to which it belongs, existing LIS programs in Taiwan up to 2011

Institute name	Location	Year established	College/school to which it belongs	Existing programs and year established
NTU	Taipei	1961	College of Liberal Arts	UG—1961 G—1980 PhD—1989 LS to LIS in 1998
SHU	Taipei	1964	College of Journalism and Communications	ND—1964–1993 UG—1995 LIS G—2000 2001 changed name to Information & Communications[a]
FJU	Taipei	1970	College of Liberal Arts 1970–2010 College of Education 2010 to present	UG—1970 LS to LIS 1992 G—1994
TKU	Taipei	1971	College of Liberal Arts	UG—1971[b] G—1991 2000 changed name to Information & Library Science
NCCU	Taipei	1996	College of Liberal Arts	G—1996 PhD—2011
HCU	Hsin-Chu	1998	College of Information & Communication	UG—1998–2010[c]
NCU	Taichung	1999	College of Liberal Arts	G—1999
NTNU	Taipei	1954	Dept of Social Science	U—1954
		2002	College of Education	G—2002 PhD—2010
NJTU	Hsin-Chu	2002	College of Engineering and Computer Science	G—2002–2010[d]

Source: 2012 Librarianship Yearbook of LAROC Taiwan, pp 207–235

Abbreviations of Institution Names (Department/University Names) in Table 6.1

NTU Department and Graduate Institute of Library and Information Science, National Taiwan University, *SHU* Department and Graduate Program of Information and Communications, Shih Hsin University, *FUJ* Department and Graduate Program of Library and Information Science, Fu-Jen Catholic University, *TKU* Department and Graduate Institute of Information and Library Science, Tamkang University, *NCCU* Graduate Institute of Library, Information and Archival Studies National Chengchi University, *HCU* Department of Library and Information Science, Hsuan Chuang University, *NCHU* Graduate Institute of Library and Information Studies, National Chung-Hsing University, *NTNU* Graduate Institute of Library and Information Studies, National Taiwan Normal University

Librarianship Education Program under Department of Social Science

NJTU A Digital Library and Information Degree Program of ECE and CS Colleges (College of Engineering and Computer Science), National Jiao-Tong University, *UG* Under Graduate Program, *G* Graduate Program, *PhD* Ph.D. Program, *ND* Non-Degree Community College Program

[a]Department renamed to Information and Communications in 2001

[b]Department renamed to Department of Information and Library Science in 2000

[c]Department was merged with College of Information & Communication in 2010 and stopped recruiting students since then

[d]This program stopped recruiting students since 2010

Ph.D. Programs, NTU established the first LIS Ph.D. Degree Program in 1989. 20 years later, NTNU and NCCU also obtained permission to start the Ph.D. Programs in 2010 and 2011 respectively. With the approval of two additional Ph.D. programs have broaden the scope of professional development, The existing three Ph.D. programs distinguish the specialization in traditional LIS curriculum (NTU), teachers' librarianship (NTNU), and archival studies in the field (NCCU).

6.3 Quality Assurance and Quality Assurance System Defined

6.3.1 Quality Assurance Defined

Based on the Oxford Dictionaries (2013), quality is defined as the standard of something as measured against other things of a similar kind and the degree of excellence of something. Council for Higher Education Accreditation (CHEA) defined quality as the suitability of a target with the specification. Wise Geek (2013), a team of writer, researchers and editors in United States found that quality assurance is a procedure that focuses on enhancing the process that is used to create the end result. It is not emphasized the result itself, quality assurance is a certain form of ordering and naming to ensure the quality. It is a mode on measuring, improving, and maintaining the quality of any valued human activity included education, academic, manufacturing, health care, service, business, sports, infrastructure, or governance. It is all involved a planned system but not directly in a development process. Adebayo (2009) defined quality assurance as "a means of ensuring that the best practices are encouraged in a social system. Quality assurance refers to the systematic activities implemented in a quality system so that quality requirements for a product or service will be fulfilled. Quality assurance always helps to clarify the measurement of standards in subject via subject benchmarking groups. Also, provide a greater public assurance. As a result, quality assurance principles are as indicators in accordance to comply with.

6.3.2 Quality Assurance in Education

Four reasons identified by Ekhaguere (2006) on implementing quality assurance: (1) Quality assurance used as cutting cost on reducing the poor quality outcome; (2) Human normally prefers good quality products; (3) Good quality is showing the capability. Poor education rise poor economy; (4) Quality assurance process involves setting up standards and ensuring that the standards established are kept to, and reviewed periodically.

During the last decade or two, many countries have created their own quality assurance mechanisms. Thus, International Institute for Educational Planning/UNESCO stated that education with quality assurance has become an important global trend. Quality of education is decided by consumer based on the suitability, correctness of an educational product. Educators have to develop a specification that can fulfill consumer needs and make the greatest effort to reach the standard. Quality assurance played an important role on ensuring the acceptable standards of infrastructure, education and scholarship are being maintained and enhanced. (Council for Higher Education Accreditation, CHEA, 2003).

6.4   Quality Assurance System of LIS Education in Taiwan

6.4.1   Goals and Objectives of LIS Education in Various Levels

6.4.1.1   Goals/Objectives for Undergraduate Level: Bachelor's Degree

Goals and objectives of LIS Education for undergraduate level, i.e., for B.A. Degree, were accumulated as follow:

1. Offering the knowledge and training in the field of library and information science.
2. Fostering library personnel with the ability to think independently.
3. Conducting research and offer services.
4. Nurturing team spirit and good coordination and communication skills.
5. Building up diversified value and international scope and vision.
6. Consolidating the administration and teaching of leadership.
7. Designed to train students with the capability of collecting, managing, evaluating.
8. Disseminating information and knowledge to become LIS professionals.

6.4.1.2   Goals/Objectives for Graduate Level: Master's Degree and Ph.D. Degree

1. Goals/Objectives of Master Degree Programs
   The goals and objectives of Master's degree level including what follows:

   (a) Cultivate mid-range professional leadership of information service institutes and enabling students to be equipped to conduct services in the information service professions;
   (b) Educating students to conduct research and enabling students to be able to enhance quality and quantity of research in the library and information service professions;

(c) To educate students with the knowledge and capability of LIS theory and practice, as well as related theories and practical information technology;

(d) Offering the knowledge and training in the field of library and information science;

(e) Fostering library personnel with the ability to think independently make research and offer services;

(f) Nurturing team spirit and good coordination and communication skills;

(g) Building up diversified value and international scope and vision;

(h) Consolidating the administration and teaching of leadership.

2. Goals/Objectives of Doctoral Degree Programs

These institutes stand on educating professionals with further depth academically to address:

(a) The nature of information and its uses;

(b) Supporting technologies through teaching, research, service, and leadership;

(c) Educating and training higher level professionals to enhance the capability of the field as well as the innovative ability for the profession;

(d) Educating and training teaching forces as well as lifting research capability for library and information studies.

Curriculum design, in general, of LIS education focus on meeting the diversified needs of information services. To fulfill the services, students are expected to obtain the following capabilities:

1. Producing, communicating, and utilizing information resources.
2. Skills of information retrieval and access.
3. Information and knowledge organization.
4. Information service—of professionalism and communication skills.
5. User research—understanding user needs of different types of library.
6. Information resource management and knowledge management.
7. Information analysis and study.
8. Management of libraries and information centers.
9. Information and Communication Technologies.
10. Evaluating information and library utilizing.
11. Knowledge innovation—humanity and creation of knowledge and information.

6.4.2   Requirements of LIS Education Degrees in Various Levels

6.4.2.1   Teaching Qualifications

Entry level of teaching in the university requires Ph.D. Degree (exception on those entered in before year of 2000) as Assistant Professor, fulfilling with at least 3-year teaching and research outcome with publication, passing through review process to become an Associate Professor. Qualified Associate Professors go through at least 3-year of teaching and research and services. At least a formal publication (book) is a must, plus qualified papers published on academic journals. A formal process of examination and further review steps to certify an Associate Professor to become a full Professor.

Teachers are reviewed periodically by authority. The process is usually done by the university which forms up a committee to examine categories of teaching, research, and services by committee members. Certificates of Lecturer, Assistance Professor, Associate Professor, and Professor are issued by Ministry of Education via application of individual university.

During 1970–1990, about 70 % of the teaching force was from teachers with foreign degrees, mostly from the USA. However, in the recent 20 years, it has gradually turned into 50 % due to availability of local scholars with doctorate degree. This will becoming more obvious in the next 6 years when the two new doctoral programs were permitted in 2010 and 2011, soon new faculty member with local Ph.D. degrees will join the teaching group. One can identify that the ecology of the profession will be different when the portion of scholars with foreign degrees and local degrees changes. The content of teaching material, the direction of professional development, and philosophy of the education will be different as well.

6.4.2.2 Degree Earning Requirement

1. Requirement of Undergraduate Level for B.A. Degree
   To earn a B.A. Degree from a LIS Program, taking NTU's undergraduate program as an example, a student must obtain a minimum of 139 credits of course work including 30 credits required by the University, 61 required and 28 elective credits for the major subject specialty, and 20 elective credits from other departments. Students need to earn 60 points and above out of 100 points to earn credit of a course. Students may repeat the course till passing the requirement. Students will be dismissed by the university if failed twice to earn 50 % of registered credits. Required credits vary in different universities, from 128 to 142 credits, so as the number of required and elective credits.
2. Requirement of Graduate Level for Master's Degree
   To earn an M.A. degree, example from NTU's LIS Graduate Program, a student must obtain a minimum of 30 credits of course work, including 12 credits required by the University, 18 elective credits for the major subject specialty, Students are allowed to take courses from other departments (6 credits maximum) with permission of the Director or adviser of the Graduate Program. Students are required to complete a thesis (0 credits, required) with academic quality which is examined by professors from outside and within the university. 70 Points (or B minus) is a minimum score to earn the credit, scores below 70 is considered a failure. Students will be required to repeat the course with additional tuition fee. Other universities may require 3-credit worth for thesis guidance.

Table 6.2 Summary of various degree requirements

Degree type	Requirements				
	Credits score	Thesis	Time-span	Presentation or publication articles	Remarks
B.A.	128–134 60 and above	No	4 years	No	Library practicum
M.A.	30 70 and above	Yes	3–5 years	No	
Ph.D.	24 70 and above	Yes	2–7 years	Yes	Qualify exam Dissertation defense exam

Students whose domain knowledge, i.e., undergraduate majors, is not LIS need to take additional 4 courses of "Introduction to Library and Information," "Information Collection and Organization," "Reference Resources and services," and "Library Practicum" as Prerequisite and complete at the first year. Students need to complete the required course at the first and the second year of the program. These clauses are not applied to LIS Programs of other universities. However, the essence of the requirement is similar among LIS Programs.

3. Requirements of Graduate Level for Ph.D. Degree
   To earn a Ph.D. degree, example from NTU's LIS Graduate Program, students are required to complete the following items:

   (a) The study to be completed within 2–7 years.
   (b) With achievement of required credits.
   (c) Pass the Qualify Examination.
   (d) Presenting a qualified paper in a national or international conference (include poster presentation in an international conference).
   (e) Publish papers, one in Chinese in a selected journal (Taiwan Social Science Index or Taiwan Human Cultural Index Core) and one in English in an English language journal.
   (f) Participating professional and academic activities.
   (g) Pass the Degree Seeking Examination.
   (h) Complete a qualified dissertation.
   (i) Pass Moral Score every Semester.
   (j) Complete required credits.
   (k) Achieve additional items, if required.

Credit number requirement is 24, degree dissertation is required but with 0 credits. Students with MA in LIS should complete the 24 credits in 2 years. Others should complete in 4 years for those whose MA is not in LIS field. Students will be dismissed if they could not fulfill above conditions. Students may take course outside campus, with 6 credits maximum with permission of the Dean or the Course Committee. The summary of requirements on credits, scores, thesis, time-span, presentation/articles of various level of degrees as Table 6.2.

6.5 Discussion

6.5.1 Stages and Practices of Quality Assurance Mechanism

6.5.1.1 Role of Ministry of Education

LIS Education in Taiwan can be categorized into three stages,

1. 1954–1979 was the initial stage.
2. 1980–1990 was the developing stage.
3. 1991 to present is an innovating stage. The LIS Education quality assurance mechanism also came along with the milestones such as: 1954–1979, the Ministry of Education (MOE) strongly supervised the higher education.

Credits requirement, curriculum design, and student recruitment of LIS education strictly followed the decisions and policies of MOE; (2) 1980–1990 was the developing stage; the concept of "Professors Administrate Universities" was introduced and universities became more independent in managing the university. MOE played a role of archiving and recognition; (3) 1991 to present, LIS Education has been actively promoting their curriculum and faculty development to enhance the contents of teaching and learning. MOE plays a role of further recognition and supporting, such as awarding "Excellent Prize" to those universities prepared projects to improve teaching and learning.

6.5.1.2 Stages and Emphasis of LIS Education in Taiwan

As shown in the Fig. 6.1 of Stages and Emphasis of LIS Education in Taiwan, Lin (Lin, 2007). At the Stage I when LIS education was titled "Library Science" and "Library & Information Science" the emphasis was at the contents of library holding, including collection development, acquisition, cataloging of information. At the stage II, LIS education was emphasized on information creation, i.e., information formation, contents digitizing, topic (subject) database, data warehousing in addition to contents acquisition. At the Stage III, LIS education has been emphasizing on innovative information services, the information and communication technologies has helped the librarianship moved into another level of information services.

Figure 6.1 identifies that the information services have been transformed from contents to information consumers.

6.5.1.3 Committee on Education, Library Association of Republic of China, Taiwan (LAROC)

Faculty, Students, and Environment play influential roles in teaching, learning, and cultivation of the professional education. The development of the LIS Education in Taiwan has achieved a great outcome in the last 50 years. However, the entrance of

(stages)

Stage III
Transforming *(Sharing)*
IC: Information Communication

Stage II
(Creating)
I: Information

Stage I
(Acquiring)
LI: Library & Information

(emphasis)   ***Contents***   ***Consumers***

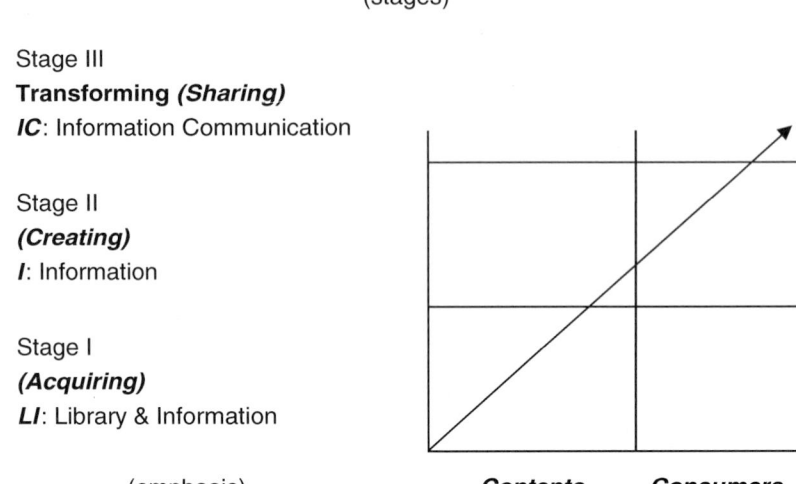

Fig. 6.1 Stages and Emphasis of LIS Education in Taiwan: *Source*: Lin, Chihfeng P. (2007) "Pillars of Educational Foundation for Information Services Professionals" Proceeding of the second International Conference of Asia-Pacific Library & Information Education & Practices (second A-LIEP 2007)

librarianship was required by civil examination or by contract-base employment, thus the degree does not guarantee the employment of libraries. At the meantime, librarians are not necessary LIS graduates in Taiwan.

This has been especially true in counties outside the major cities, public libraries employed workers who did not have LIS education background. To deal with fact that libraries need professional services, LAROC provided on-the-job training projects of subject-basis which is short term, usually 1-week course to train these workers to become more professional in library services. These projects have been continued and updated the contents to go along with the advancement of LIS development since the last 30 years.

6.5.1.4 Library Law Accomplished in 2001 After 40 Years of Efforts

A 20–clause Library Law of Taiwan accomplished and announced in 2001. This is an efforts outcome of 40 years by LIS profession. The Library Law indicates categories of library, administrative level, qualification of library directors of each type, and duties of each type of library and their services. It is an initial base law of the kind, the related policies and regulations have been undergoing and some were announced. This Library Law gives indications to librarianship. However, as soon it was issued, it is dated due to the change of information environment. The LIS profession is now working on the revision and trying to update the contents to meet the current needs.

6.5.2 Quality Assurance System as Indicator of Common Awareness

6.5.2.1 Students Evaluating Teaching Quality

Universities now also conduct "Teaching Quality Assurance Evaluation" at the midterm of the Semester and the final of the Semester. Example of SHU, the evaluation criteria include teaching materials, teaching methodologies, and teaching ethics. Students give score from 1 point (not good) to 5 points (very good), if the average score is below three points, full-time teachers need to take advices from the director of the department. If the adjourned teachers get average score below 3 points, they will not be invited again in the next Semester.

6.5.2.2 Evaluation of University/College Nationwide

1. First-Run (2006–2010) Evaluation Items of Departments of University/College Nationwide
   "Higher Education Evaluation & Accreditation Council of Taiwan," a cooperative agency was initiated and entrusted to examine each university's Education Goals and Curriculum Design, Resources of Teaching and Learning, Professional Performance of Faculty Members, Learning outcome of Students, Administrative Effects, Control, Improvement, and Development.
   The items, Contents, and Indicators include:

   (a) Objectives, Specialty, and Self-Improvement
   (b) Curriculum Design and Teaching of Faculty
   (c) Student Learning and Student Affairs
   (d) Research and Professional Performance
   (e) Performance of Graduates
      A systematic process was designed and carried out of each department of every university. Each department should form up an ad-hoc committee for the evaluation, usually all faculty members are encouraged to get involved. A prior evaluation practice should be conducted, scholar(s) from outside the school were invited to do the evaluation and provide comments and suggestions before the formal one does. The factors and efforts of improvements in the prior evaluation will be included in the formal evaluation process. Each and every department of universities went through the similar process. Most

departments passed the examination, some were under "probation" (department/
university can request reevaluation optionally), and few resulted to suspend
the recruitment included one in the LIS education.
2. Second-Run (2012–2016) Evaluation Items of Departments of University/
   College Nationwide
   Six years later after the first run of nationwide evaluation, the second run is
   underway. The contents of evaluation include:

   (a) Objectives, Core Competency, and Curriculum Design
   (b) Teachers' Teaching and Learning Measurement
   (c) Guidance and Counseling and Learning Resources
   (d) Academic and Professional Performance
   (e) Performance of Graduates and Self-Improving of whole

The second-run evaluation emphasizes on self-evaluation of department/college.
It is also emphasizes on student-centered assessment, especially to reveal that if the
learning meets the job market. The evaluation process has been simplified as to
invite specialists of the field to review the self-evaluation documents, discuss with
the faculty member on-site, revise the evaluation document and submit the documents to the "Higher Education Evaluation & Accreditation Council of Taiwan"
and MOE. Specialized committee will review the documents and the process to
instruct if the evaluation will be passed.

Cross-country education development has been going for a long period of time.
Institutes of higher education in countries are providing opportunities for students
to learn across the borderlines as well expansion of globalization. This also brought
in issue of quality assurance in higher education.

6.6 Conclusion and Suggestions

Synthesizing of Library, Information, Communication, Archival Studies, Quality
Assurance System of LIS Education in Taiwan can divided into stages as the profession emphasized in library science, library & information science, and information
and communications. Role of Authority—Ministry of Education (MOE) had become
as supportive role and LIS education could work independently in curriculum design
and faculty development. The stages and practices of quality assurance mechanism
have been concrete with base of Library Law. Even though Committee on Education
of LAROC does not stand strong on the evaluation mechanism, but it has been helpful to strengthen the LIS profession by providing training projects. LAROC may
become an important unit to help evaluate the LIS education if the process can be
put on track of professional development. MOE is also preparing to organize a
Committee on Librarianship which is resumed from the same unit two decades ago.

"Higher Education Evaluation & Accreditation Council of Taiwan" plays a vital
role of quality assurance with emphasis on Objectives, Specialty, and Self-
Improvement 6 years ago. The examination movement reshaped the ecology of

Taiwan's higher education, as well as LIS education. One program was diminished and further changes are expected. Continued with another run of examination, with emphasis on Objectives, Core Competency and Curriculum Design, and Performance of Graduates and Self-Improving of whole are underway. Eventually, the direction of improvement will be counting on accreditation unit of professional organization, or at least cooperating with the professional organization to execute the evaluation process. In this case, the connectivity with international professional organization such as IFLA, ALA, and ASIS&T become important. To acquire the trends and future perspectives of particular sections, divisions, or committees related to LIS education, such as Section of Education and Training (SET), Section of Continuing Professional Development at Working Place (CPDWL), and Committee on Standards of IFLA, Committee on Education of ALA, and Committee on Education of ASIS&T. International professional organizations should work together to improve the quality of LIS education as their common goals.

To establish, and Taiwan already has one, an independent organization of higher education evaluation for quality assurance is an outcome of being a member of International Network for Quality Assurance Agencies in Higher Education (INQAAHE). A recent accomplished "2011 HEEACT International Conference—Internationalization of Standards in Higher Education: Accountability, Student Learning Outcomes and Collaborations in Quality Assurance Agencies" stated regional influence on international issues in higher education. This also reflect to LIS education that the connection of libraries with other information service institutes, such as museums, archives, records, and content curator (or galleries) ought to be linked and expand their partnership with The World Intellectual Property Organization (WIPO), International Standards Organization (ISO), International Publisher's Association (IPA), and International Federation of Library Associations and Institutes (IFLA). These activities could not be accomplished by individual institute rather than cooperation and collaboration regionally, internationally. LIS education in Taiwan has been achieved outstandingly and a road ahead awaits further development.

Appendix 1: LIS Education Institutes in Taiwan and URL

Department and Graduate Institute of Library and Information Science, NTU http://www.lis.ntu.edu.tw/

Graduate Institute of Library, Information and Archival Studies, NCCU http://www.lias.nccu.edu.tw/

Graduate Institute of Library and Information Studies, NTNU http://www.glis.ntnu.edu.tw/webpage/

Graduate Institute of Library and Information Studies, NCHU http://www.gilis.nchu.edu.tw/

Degree Program of ECE and CS Colleges, NCTU
http://dpeecs.nctu.edu.tw/introduction/index.aspx

Department of Library and Information Science, Fu Jen Catholic University, FJU
http://web.lins.fju.edu.tw/drupal/

Department and Graduate Program of Information and Communications, SHU
    http://ic.shu.edu.tw/

Department of Information and Library Science, TKU
http://www.dils.tku.edu.tw/

Department of Library and Information Science, HCU
http://dlis.hcu.edu.tw/front/bin/home.phtml

References

Adebayo E. L. (2009). Quality assurance and the implication for the management of university libraries in nigeria—university librarian federal university of technology, Akure. *Library Philosophy and Practice,* ISSN 1522-0222
Council for Higher Education Accreditation. (2003). CHEA.
HEEACT www.heeact.edu.tw/ct.asp
Lin, C. P. (2007). *Pillars of educational foundation for information services professionals.* Proceeding of the Second International Conference of Asia-Pacific Library and Information Education and Practices (Second A-LIEP 2007).
What is Quality Assurance Wise GEEK clear answers for common questions. May 31st 2013 Retrieved from http://www.wisegeek.com/what-is-quality-assurance.htm

Chapter 7
LIS Education and Quality Assurance System in Asia-Pacific: Malaysia

Mohd Sharif Mohd Saad[†], Rusnah Johare, and Fuziah Mohd Nadzar

7.1   Introduction

Throughout the history of human civilization, libraries and librarians have been known to be catalysts for intellectual growth and lifelong learning among individuals; strategic players for corporations in making business and managerial decisions; gatekeepers and knowledge managers for policy makers and governments of nations; providers of knowledge for academics and researchers to further their intellectual pursuit and new discoveries; and teachers who inculcate reading habits among young children (Szarina, 2002). Hence, the efficiency and effectiveness of information and library services of a nation is proportional to the status of the nation's development. As mentioned by the former Malaysian Prime Minister Tun Dr Mahathir Mohamad (1991) in the context of Malaysian Vision 2020 and the concept of knowledge-based economy : "There is no rich country that is information-poor. And there is no poor country that is information-rich." Keeping pace with the Vision 2020 and the implementation of the concept of knowledge-based economy, five universities in Malaysia have placed a strategic importance to LIS schools. These are Universiti Teknologi MARA (UiTM), University of Malaya (UM); International Islamic University (IIUM), University of Selangor (UNISEL) and Islamic Science University of Malaysia (USIM). The practice of quality management in LIS education in these five universities is governed by the Malaysian government quality assurance policies.

M.S.M. Saad (Deceased) • R. Johare (⊠) • F.M. Nadzar
Faculty of Information Management, Universiti Tekonogi MARA (UiTM),
Shah Alam, Selangor, Malaysia
e-mail: r.johare@yahoo.com; fuziahmn@salam.uitm.edu.my

M. Miwa and S. Miyahara (eds.), *Quality Assurance in LIS Education:*
An International and Comparative Study, DOI 10.1007/978-1-4614-6495-2_7,
© Springer Science+Business Media New York 2015

7.2 History of Library and Information Studies Education in Malaysia

The beginning of library and information science education is always associated with the formation of the Malaya Library Group (MLG) in 1955 which is at present known as the Librarians Association of Malaysia (PPM) (Lim, 1970). Shortly after Malaya gained her independence in 1957, the association initiated to promote the profession, mainly in the areas of professional training, education, and qualification of librarians (Wijasuriya, Lim, & Nadarajah, 1975). In the early years, MLG organized library and information studies classes solely to enhance the quality of library services in Malaya (Abdoulaye, 2004; Edzan & Abdullah, 2003). Grants from the Asia Foundation and Smith-Mundt enabled the conduct of many short courses despite the unavailability of a library school at that time (Nadzar, Sidek, & Saad, 1993). Classes in librarianship organized in the 1960s then were short vacation courses for teacher librarians and formal courses for students sitting for the UK Library Association examinations. The British LA courses were at that time offered by Institut Teknologi MARA (ITM) from 1968 to 1971. Candidates need not be in the UK to pursue a professional library qualification.

However, when the Library Association of UK discontinued offering the external library qualifications to countries outside the UK, the School of Library Science, ITM, consequently established a 3-year Diploma in Library Science program (equivalent to a General Degree qualification). The priority at that point of time was to produce undergraduate qualified professionals. At its initial phase, the librarianship program was replicated from the ALA syllabus, and was later redesigned and embedded with local contents (Saad & Seman 1995). In 1972, ITM started its Postgraduate Diploma in Library Science to cater for university graduates who wish to gain employment in library services. University of Malaya officially launched the Master of Library and Information Science program (MLIS) during their 1987/1988 session under the auspices of the Institute of Advanced Studies. However, it was suspended the following year, but was then revived in November 1994 and hosted at the Faculty of Computer Science and Information Technology.

Over the years, a number of universities offered LIS courses at both the undergraduate and postgraduate levels. The International Islamic University (IIUM) started offering its Masters in Library Science and Information Science program (MLIS) since 1992. The Department of Library and Information Science is at the Kulliyah of Information, Communication and Technology (ICT). Universiti Teknologi MARA started its Masters of Science in Information Management in 1997. Subsequently, this was followed by the Master in Library Science program (MLS) in 2006. In 2003, the Islamic Science University of Malaysia (USIM) commenced the Bachelor of Sunnah Studies with Information Management. In recent years, University of Selangor (UNISEL) established its Diploma in Library Science program in 1996, and in September 2011 they established the Bachelor in Library Science program. A milestone of the historical development of library and information studies in Malaysia is depicted in Table 7.1 below.

Table 7.1 Milestones in the development of library and information studies programs in Malaysia

Year	Milestone
1960	W. J. Plume, University of Malaya (UM) Librarian submitted a proposal to UM authorities to include a library school in the second phase of the university library programs. It was supported by Persatuan Perpustakaan Malaysia (PPM)
1960s	PPM conducted part-time classes to prepare candidates for the Library Association Examination, UK
1965	PPM sent a memorandum to the UM authorities urging the establishment of a library school
1967	The Higher Education Planning Committee (HEPC) report revitalized the establishment of the library school
1968	Institut Teknologi MARA (ITM), established a Department of Library Science under the School of Public Administration and Law. It conducted a full time program in Librarianship preparing students for the British Associate of the Library Association (ALA, UK)
1969	At the International Council of Archives (SARBICA) Conference in Jakarta, both SARBICA and PPM agreed to the establishment of a postgraduate school of Librarianship at UM
1970	ITM established the School of Library Science; changed its name to "School of Library and Information Science" in 1979; it was once again changed to "Faculty of Information Studies" in 1997; and the current name "Faculty of Information Management" has been in use since 2005
1972	A memorandum was sent to the National Library Committee (NLC) for the establishment of the graduate school at UM
1972	With the end of the external ALA program, ITM introduced a 3-year Diploma in Library Science program, with local contents planned into the curriculum
1972	ITM introduced a 1-year Postgraduate Diploma in Library Science program
1986	The first National manpower survey of libraries and information services in Malaysia was conducted by ITM and UNESCO
1987	The Masters in Library and Information Science program (MLIS) was offered by UM for the 1987/1988 session. However, the course was suspended the following year
1991	ITM's 3-year Diploma in Library Science (equivalent to the General Degree) was upgraded to a 4-year Honors Degree program; and in 1999, ITM became Universiti Teknologi MARA (UiTM)
1992	The International Islamic University, Malaysia (IIUM) introduced the MLIS program at the Department of Library and Information Science, Kulliyah of ICT
1994	The MLIS program at UM was revived and housed at the Faculty of Computer Science and Information Technology
1996	University Selangor (UNISEL) launched the Diploma in Library Science program
1999	Universiti Sains Islam Malaysia (USIM) started a Bachelor degree in Sunnah Studies with Information Management
2004	A study on human resources need for Library and Information Services in Malaysia (Zaiton et al.) commission by The National Library of Malaysia
2006	UiTM started the Master in Library Science program
2011	UNISEL started the Bachelor in Library Science program

7.3   Background of Malaysian Quality Assurance Policies

Under the Malaysian law, all institutions of higher education established by the federal government are deemed to be self-accredited. The first self-accrediting institution in Malaysia is the University of Malaya (UM), the first university in the country established by the University of Malaya Ordinance 1949. The Ordinance empowered UM to confer diplomas and degrees in its own name. To re-affirm the authority granted under the UM Ordinance 1949, two new legislations were introduced, namely, the University of Malaya Act 1961 and the Degrees and Diplomas Acts 1962, when UM was reorganized in 1961.

In 1971, the Malaysian federal government granted all its universities and university colleges the power to confer diplomas and degrees in their own name through the Universities and University Colleges Act 1971 (UUCA). Pursuance to this development, Institut Teknologi MARA (ITM) which was established in 1956 under the name of Maktab Latihan RIDA (RIDA Training College) was granted the same power through ITM Act 1976. In 1997 ITM was granted a university status granting it the same powers as other institutions established under the UUCA.

As far as privately owned institutions (including owned by state governments) are concerned, no formal accreditation system for courses of study existed until the passing of a series of education-related legislations in 1996. The subsequent passages of the Private Higher Educational Institute Act 1996 and the National Accreditation Board (Lembaga Akreditasi Negara) or LAN established later in the same year saw the revamping of the registration and approval regime of private institutions of higher education in Malaysia. Consequently in April 2002, a Quality Assurance Division (QAD) was established by the Ministry of Education (MOE) to manage and coordinate the quality assurance system in public universities.

7.4 The Malaysian Qualifications Agency (MQA) and Malaysian Qualification Framework (MQF)

In June 2003, a national consultation seminar was held to establish a national qualification framework (Malaysian Qualification Framework- MQF) that would integrate, rationalize, justify and bring together all qualifications offered on a national basis into a single interconnected system (Ministry of Higher Education, Malaysia: QAD—Position Paper for a National Qualifications Framework—http:apps.emoe. gov.my/qad/nqf/PP.doc) (Table 7.2).

The MQF was finally adopted in 2007 with the establishment of the Malaysian Qualifications Agency (MQA), a statutory body set up under the Malaysian Qualifications Act 2007 (Ministry of Higher Education, Malaysia, 2008). The QAD and LAN were dissolved and their functions were taken over by the MQA (MQF at a Glance—http:mqa.gov.my/index1.cfm).

The main role of the MQA is to implement the Malaysian Qualifications Framework (MQF) as a basis for quality assurance of higher education and as the reference point for the criteria and standards for national qualifications through the following functions (MQA, http//www.mqa.gov.my):

• To implement MQF as reference point for Malaysian qualifications. To develop standards and credits and all other relevant instruments as national references for the conferment of awards with the cooperation of stakeholders

Table 7.2 MQF qualification and levels chart

| MQF levels | Sectors | | | |
	Skills	Vocational and training	Higher education	Lifelong learning
8			Doctoral Degree	Accreditation for Prior Experiential Learning (APEL)
7			Master's Degree	
6			Bachelor's Degree	
			Graduate Certificate & Diploma	
5	Skills Advanced Diploma	Advanced Diploma	Advanced Diploma	
4	Skills Diploma	Diploma	Diploma	
3	Skills Certificate 3	Vocational & Technical Certificate	Certificate	
2	Skills Certificate 2			
1	Skills Certificate 1			

Source: Malaysian Qualifications Framework (http:www.mqa.gov.my)

- To quality-assure higher education institutions and programs
- To accredit courses that fulfill the set criteria and standards
- To facilitate the recognition and articulation of qualifications
- To maintain the Malaysian Qualifications register (MQR)
- To evaluate foreign qualifications and assess it for equivalency with Malaysian secondary school and university preparatory qualifications.

7.5 The Application of Quality Assurance in Malaysian LIS Education

In Malaysia, the government plays an active role in defining the pace of development and the maintenance of quality in the public and private education industry through the establishment of the MQA. The MQA governs the standards and quality of higher education offered by the Public Higher Education Institutions and Private Higher Education Institutions in the country. Programs need to be approved by MQA and conform to MQA's standards before they can be offered. Thus to conform to the requirements of MQA, most higher education, private or public, have a set of quality assurance measures to safeguard quality and reputation. Even though each of the five universities offering LIS education has a different purpose, distinct expectations, varying motivation and different efforts on quality assurance, their common concern is to plan a systematic review process of their LIS programs to determine that acceptable standards of LIS education, scholarship, and infrastructure are being maintained and enhanced.

7.6   UiTM's Experience in Implementing Quality Assurance for LIS Education

7.6.1   Quality Assurance in the Early Years

Realizing the important role of libraries and the library profession, the government of Malaysia saw it necessary to establish the first formal education program for professional librarians in Institut Teknologi MARA (ITM), now Universiti Teknologi MARA (UiTM) in 1968. This was alongside other academic courses for professional accountants, architects, engineers, lawyers and secretaries. It was without doubt that graduates of these early courses have become pioneers in their respective professions, providing leadership and foundation for younger generations to build upon.

The Library Science program was established initially under the School of Administration and Law in 1968 to prepare students for the professional examination of the Library Association (UK), so as to enable them to qualify for the Associate ship of Library Association (A.L.A.). In the early years of the School establishment (1968–1971), the A.L.A. qualification of the UK professional body was recognized by the Malaysian Government, and holders of this A.L.A. gained employment as professional librarians or are posted at the equivalent level of government officers holding a Bachelor with Honors degree from any recognized university.

When the School started its own 3 year Diploma in Library Science program in 1972 and admitted students with a Higher School Certificate, it placed the program at a par with other Bachelor's degree programs at Malaysian universities. For quality assurance the Diploma then needed to be recognized by the government so as to enable graduates to secure employment at certain acceptable levels. The School's management under the leadership of successive heads made consistent efforts to design the curriculum to be at par with those conducted at other universities, in terms of entry requirements, contents and length of the program. Preparation of documents to convince the Public Services Department of Malaysia (PSD) on the merit of each program and quality of its graduates became one of the processes of quality assurance mechanism then. For this purpose, all diploma programs were scrutinized and the graduates' performance examined on the job before the recognition became officially endorsed. In February 1981, the official recognition was bestowed upon the first two programs, namely, Diploma in Library Science (Full-time 3-year Program), recognized by the PSD as equivalent to a general Degree), and Postgraduate Diploma in Library Science (Full-time 1-year Program), recognized by the PSD as equivalent to a professional qualification in addition to a Bachelor Honors Degree.

Later, in July 1994, the Advanced Diploma in Library Science or Bachelor of Library Science (Hons.) (Full-time 4-year Program), was recognized by the PSD as equivalent to a Bachelor with Honors Degree. When the Institute's status was changed to that of a university on 26 August 1999 (its name was changed to Faculty of Information Studies, and later renamed Faculty of Information Management, henceforth referred as FIM), it automatically conferred appropriate degrees on its graduates. The challenge then was for the Faculty to obtain recognition for each program at the international level in order to enhance the global competitiveness of the programs and the graduates. These are programs with four specializations, namely, Bachelor in Information Studies (Hons.) (Library and Information Management); Bachelor in Information Studies (Hons.) (Records Management); Bachelor in Information Studies (Hons.) (Information Systems Management); and Bachelor in Information Studies (Hons.) (Resource Center Management). These specialized programs were introduced to establish a strong knowledge base in various areas related to the field of library and information science, including archives and records management, information systems management and resource center management. To fulfill the requirements of quality assurance systems of the university and that of the information industry, students are provided with the necessary skills and competencies enabling them to perform at the initial level of a professional career in libraries and information-related organizations. The Faculty then has also started a new program, Master of Science in Information Management.

In August 1999, another historic moment took place, when ITM was given a university status by the Prime Minister of Malaysia and had a name change to Universiti Teknologi MARA (UiTM). In conjunction with this major milestone, the Faculty introduced two new programs, namely, the 3-year Diploma in Information Management offered for holders of Malaysian Examination Certificate, and at the same time a Doctoral program in Information Management. In the following year, the Faculty started the Master of Science in Information Management program via flexible learning mode (FLP), thus making it the first and only faculty in UiTM to expand its Master's program to qualified students who are scattered in distant parts of the country. In 2003, another new program, Master of Knowledge Management was introduced. During 2004–2005, the Faculty successfully prepared yet another Master's degree program, Master of Library Science that enables professional librarians to advance their knowledge to fill several vacancies at managerial levels in all types of libraries across the country.

During these years, the development of the Faculty's programs were very much shaped by the QAD requirements and also the new requirements of the Malaysian government in 2002 where all agencies in the public sectors were to implement a Quality Management System (QMS) Standard (MS ISO 9001:2000) as a tool to practice evidence-based management. Kaur (2007) suggested that increased enrollment, greater need for accountability, stringent finance and most of all to be internationally recognized, drove the education sector, especially universities and colleges, to begin to experiment with this international standard that focus on the quality and reliability of processes that create products and services. To get the certified MS ISO 9001:2000 academic university status (mandatory Malaysian government requirements) UiTM has appointed an external consultant, Lloyd's Register Quality Assurance Ltd (LRQA) to carry out an assessment of its management system. To establish compliance with the Assessment Standards, the Faculty had undergone an assessment and audit exercise by LRQA in the form of surveillance visits, document review visits and assessment. Assessment was based on systems and processes involving students' enrollment, quality teaching and learning and research activities in the form of Quality Audit Reports (QAR). Pursuance to QAR 2007, the Faculty provision of tertiary education for Bachelor of Science in Information Studies (Hons.) and Management of Diploma in Information Management program was granted certification of MSISO 9001:2000.

Since the introduction of the Faculty's own academic program in 1972, all efforts have been made to assure the quality of the curriculum and its implementation to convince the Malaysian government or the PSD of the integrity of the Faculty's programs. Hence internationally-recognized scholars in the field were selected from among well-known academics from reputable universities to serve as the Faculty's External Examiners for a minimum term of 2 years. The External Examiners comments and advice through official written reports have helped FIM to improve the credibility and quality of the programs to be at par with those of similar faculties in other universities around the world. The Faculty continues to engage an external examiner for each program as one of the most important quality assurance mechanism.

Besides the quality assurance through the process of external examiners, the Faculty also seeks to ensure the relevance and substance of the programs through the scrutiny of academic advisors. The panel members of academic advisors comprise senior professionals in the field of library science, information science, and records management. Their advice and comments regarding the curriculum and its implementation are helpful in ensuring that there is adequate balance between theory and practice. Their role in quality assurance is important to the Faculty as they are the major employers of the graduates. The panel functions as a forum for discussion and exchange of ideas regarding various matters related to the programs, courses, curriculum, manpower needs, and industry requirements. The members of the panel comprise senior and prominent librarians from each type of library and information organizations, representing the National Library of Malaysia, the National Archives of Malaysia, one each from the public libraries, special libraries and academic libraries, the Librarians Association of Malaysia, and other organizations that may be directly or indirectly related to the main fields of study. The external examiner and academic advisor remained as important components in quality assurance elements for LIS education at the Faculty even though the QAD was abolished in 2002 and quality audit by LRQA was terminated in 2010 to be replaced by Internal Quality Audit (IQA) in accordance to the Code of Practice for Institutional Audit (COPIA) requirements of UiTM.

7.6.2   The Present Quality Assurance Systems at FIM

The IQA system in UiTM consisted of two main processes: Internal Audit to be carried out by Quality Audit Unit in every Faculty and External Review carried out by Institute of Quality and Knowledge Advancement (InQKA) started in 2011. InQKA was set up to establish and maintain a robust system of quality assurance in UiTM. It is responsible for leading and coordinating quality assurance work i.e. plans, strategies and policies carried out in all the academic centers and support services and to carry out auditing of all academic centers by InQKA's own group of trained and experienced auditors. These auditors consisted of UiTM's academic and administrative staffs who are internally appointed by the Vice Chancellor as auditors.

In order to comply with the audit processes by InQKA, all faculties need to carry out their Internal Audit by the individual Faculty's Self Review Committee (SRC) prior to External Review by InQKA in accordance to a specific template called the Self Review Report (SRR) template. The External Review Panel of InQKA uses the SRR template to test the quality assurance system in place as reported. The template includes 114 benchmarks and 65 enhanced standards in COPIA. This is to enable the SRC to evaluate itself using the same template and requirements and become aware of the detailed requirements and incorporate what is necessary in planning and practice. The SRR is used among others for quality assurance policy, identifying good practices to be nominated for best practice conventions, and the annual Vice Chancellor's Excellent Award and the Prime Minister's Excellent Award at the national level, other than for quality assurance reports needed by MQA.

Through the present system, FIM as the rest of other faculties in UiTM, has to go through an assessment of nine core areas or domains as required by the COPIA. These are: Area 1: Vision, Mission, Educational Goals and Learning Outcomes. Area 2: Curriculum Design and Delivery. Area 3: Assessment of Students. Area 4: Students' Selection and Support Services. Area 5: Academic Staff. Area 6: Educational Resources. Area 7: Program Monitoring and Review. Area 8: Leadership, Governance and Administration. Area 9: Continual Quality Improvement. The assessment on all nine core areas are based on documentation and records of existing quality procedures, quality manuals, teaching portfolios, delivery of lectures, handling of academic projects, assessment of final examinations, handling of students' course works, handling of courses without final examinations among others. All of these are based on the concept of Outcome-based Education (OBE) being practiced in all academic faculties in UiTM. The InQKA panel observations are presented on the aspects of "Strengths", "Actions in Progress" and "Areas of Concern."

7.7 Conclusion

The quality assurance systems at FIM are implemented in accordance to the models provided by Tammaro (2005), namely, Program Orientation, Educational Process Orientation, and Learning Outcomes Orientation. Program orientation stresses accountability and staffing quality indicators which include attention to the use of effective procedures in teacher selection criteria (Medical Library Association, 1992; Music Library Association, 2002; and Society of American Archivists, 2002; in Tammaro, 2005). While educational process orientation is concerned on quality audits which focuses on quality control. These involve educational needs assessment, program improvement and program justification procedures which include multiple sources of evaluation based on industrial standards such as ISO 9000,

TQM. and EQM. On the other hand, learning outcomes orientation is more concerned on explicit and detailed statements of what students learn: the skills, knowledge, understanding, and abilities which LIS Schools seek to develop and then test in accordance to the OBE requirements.

The continuous External Review by InQKA has allowed FIM to revise all programs offered at the Faculty which are based on the OBE requirements as prescribed by the Academic Affairs Division of the UiTM, and simultaneously adhere to the MQF requirements of the MQA of the Malaysian Ministry of Higher Education. These LIS programs offered presently by the Faculty have been vetted by the Academic Affairs Division of UiTM and further approved by the Malaysian government through the Ministry of Higher Education as consistent with the stated learning outcomes domains under the MQF and the soft skills requirements of the Ministry of Higher Education. All the programs have incorporated the core elements of the discipline to support the programs' outcomes.

Overall Malaysia has done quite well as regards the establishment of a regulatory framework, with clear government encouragement for adherence to externally monitored quality assurance standards and systems. The way forward would be moving from external regulatory frameworks to that of institutional self-evaluation and improvement. Only with this alignment between regulation and improvement that the measures taken will truly enhance the quality of higher education and benefit the various stakeholders.

References

Abdoulaye, K. (2004). State of library and information science education in Malaysia. *Journal of Education for Library and Information Science, 45*(1), 1–14.

Edzan, N. N., & Abdullah, A. (2003). Looking back: the Master in Library and Information Science Program at the University of Malaya. *Malaysian Journal of Library and Information Science, 8*(1), 1–18.

Kaur, K. (2007). MSISO 9001:2000 Implementation in Malaysian Academic Libraries. In A. Abdullah, et al. (Eds.), *ICOLIS 2007, Kuala Lumpur:LISU, FCSIT*, (pp 201–210).

Lim, E. H. T. (1970). *Libraries in West Malaysia and Singapore: A short history.* Kuala Lumpur: University of Malaya Library.

Mohamad, M. (1991). *Wawasan 2020.* Kuala Lumpur: Government Printing Office.

Nadzar, F. M., Saad, M. S. M., & Sidek I. (1993). *Education and training of library and information personnel in Malaysia: A status report.* Bangkok:UNESCO (Project Contract No. 848.952.2)

Saad M. S. M., & Seman, N. A. (1995). *Education and training of library professionals: Trends and perspectives.* Singapore Libraries 5.

Szarina A. (Chief Editor) (2002). *Our Intellectual Capital.* Shah Alam: Centre for Graduate Studies, Universiti Teknologi MARA.

Tammaro, A. M. (2005). *Report on quality assurance models in LIS programs.* Education and Training Section, IFLA. Retrieved August 6, 2013 from http://www.ifla.org/VII/s23/index.htm

Wijasuriya, D. E. K., Edward, L. H. T., & Nadarajah, R. (1975). *The barefoot librarian: Library developments in Southeast Asia with special reference to Malaysia.* London: Bingley.

Chapter 8
LIS Education: Quality Assurance System in the Philippines

Lourdes T. David

8.1 Introduction

This paper aims to present a brief history of the development of library education in the Philippines and the subsequent events that led to the establishment of quality assurance measures in the education and practice of librarianship in the country. It discusses the provisions of the two Republic Acts, RA 6966 and RA 9246 that provide for the standardization of the Library Science courses offered by the universities and the practice of librarianship in the Philippines. It also discusses the role of the Commission on Higher Education (CHED) in regulating courses that need board examination and the roles of library associations, the Professional Regulatory Board for Librarians (PRBL), library schools, libraries, and library associations in the continuing education of librarians. All of these bodies and activities contribute to the enhancement of the knowledge and skills of librarians and ensuring quality in library education and the practice of the profession in the Philippines.

8.2 Context

Librarians in the Philippines have had a long history of low image relative to the faculty of academic and research institutions. To help raise the image of the Filipino librarian, the Philippine Librarians Association, Inc. (PLAI), worked with various groups to professionalize the library profession. The professionalization of Librarianship required the promulgation of a law that will mandate graduates of Library Science to obtain a license or certificate of registration from the Professional

L.T. David (✉)
Ateneo de Manila University, Quezon City, Philippines
e-mail: ltdavid@ateneo.edu

M. Miwa and S. Miyahara (eds.), *Quality Assurance in LIS Education:*
An International and Comparative Study, DOI 10.1007/978-1-4614-6495-2_8,
© Springer Science+Business Media New York 2015

Regulation Commission (PRC) before they can practice. After 20 long years of struggle, Republic Act 6966 or the "Philippine Librarianship Act," was passed on September 19, 1990. Under R.A. 6966, a librarian was defined as "a bona fide holder of a certificate of registration issued by the Board for Librarians in accordance with this Act" (Republic of the Philippines, 1990). The act delineated the library science degrees that will be recognized when graduates apply to take the licensure examination. It also required licensed librarians to pursue continuing professional education programs to qualify for renewal of their licenses every 3 years.

Republic Act 6966 was repealed by Republic Act 9246 or the "Philippine Librarianship Act of 2003" (Republic of the Philippines, 2004). RA 9246 further limited the degrees that qualify graduates to take the Librarians Licensure Examination (LLE) into two, namely, the Bachelor of Library and Information Science and the Master of Library and Information Science.

The requirement for continuing education programs under RA 6966 and RA 9246 are met by the library associations, library schools and libraries by offering and sponsoring seminars, workshops, conferences, and library tours. These activities are regulated by the Professional Regulation Commission (PRC) through the Continuing Professional Education Council (CPEC) of the Professional Regulatory Board for Librarians (PRBFL). More recently self accreditation has been added as another mode for the Continuing Professional Development (CPD) of Librarians.

There are no set tools to measure quality assurance of librarians in the Philippines but in addition to the required Certificate of Registration by the PRC, certain awards and recognition given by the PRC and the library associations provide some measures of quality. There are also standards set by accrediting bodies that specify qualifications of librarians in certain positions in libraries. Scholarship grants are also available for Filipino librarians who would like to pursue higher education locally or abroad.

8.3 Library Education

During the American Regime, several public libraries were established. Dr. James Alexander Robertson, then director of the Philippine Library realized the need for trained personnel to man the public libraries. In 1910, he proposed to the University of the Philippines the establishment of a library school. Four years later in 1914, the first library courses were taught at the University of the Philippines (Damaso, 1966; Sanchez, 1966). By 1919 two schools, the University of the Philippines and the Philippine Normal College were offering library science courses (Perez, 2004).

All the teachers in the first 3-year library course were Americans. Hence the education of librarians in the Philippines followed that of the USA (Chambers, 1960). The courses offered from 1914 to 1916 were designed to provide immediate training to staff who will man the public libraries. To provide more theory to the practice, Ms. Mary K. Polk, librarian of the Bureau of Science and one of the teachers of the Library Science course proposed a revision of the curriculum in 1916 (Perez, 2004). The Board of Regents of the University of the Philippines approved the 4-year curriculum for the Bachelor of Science in Library Science on January 27, 1917 (University of the Philippines, 1922).

To ensure the continuation of the offering of Library Science courses and operation of the government libraries after the repatriation of the American teachers, the "Library Science Scholarship Act or the Alonzo Act" (Act no. 2746) was passed on February 18, 1918. The act provided for government scholars or "pensionados" to be sent to the USA for further studies in bibliography and library science from 1920 to 1923 (Philippine Islands, 1918). By 1923, seven scholars had completed their training in the USA.

Between 1914 and 1945, 11 schools offered programs in library science. All of these schools however were closed down during the Japanese occupation. After the war the schools reopened and by 2003 more than 140 schools were offering a variety of undergraduate programs in Library Science. Among these were: (1) Bachelor of Science in Library Science; (2) Bachelor of Library Science; (3) Bachelor of Science in Education with major or minor in library science; (4) Bachelor of Secondary Education major or minor in Library Science; (5) Bachelor of Arts major in Library Science; (6) Bachelor of Science in Elementary Education with specialization in Library Science; (7) Bachelor of Arts minor in Library Science; (8) Bachelor of/in Library Science; (9) Bachelor of Science in Elementary Education major in Library Science; (10) Bachelor of Elementary Education with specialization in Library Science; (11) Bachelor in Library Arts; and (12) Bachelor of Arts major in Information or Library Science. Two certificate courses, namely, Certificate in Library Service and Certificate in Library Science were also offered (Perez, 2004).

Graduate education in Library Science Started in 1952. By 2003 more than 30 schools were offering a variety of graduate programs in Library Science. Among these were: (1) Master of Arts major in Library Science; (2) Master of Education Major in Library Science; (3) Master of Science in Library Science; (4) Master of Arts in Teaching with concentration on Library Science; (5) Master in/of Library Science; (6) Master of Arts in Library Science; (7) Master of Arts in Education with Specialization in Library Science; (8) Master of Arts in Teaching major in Library Science; (9) Master of Arts in Education with specialization in School Library Science; (10) Master of Arts in Education minor in Library Science; (11) Master of Arts in Education major in Library and Information Management; (12) Master of Education major in Library and Information Management; (13) Master in Library Administration; (14) Master of Arts in Education major in Library Administration, Master of Education major in library administration; and (15) Master of Theology with specialization in Theology Librarianship (Perez, 2004).

8.4 Professionalization of Librarians

A series of legislations provided for quality assurance in the education of librarians and the practice of librarianship in the Philippines. With the passage of Republic Act 6966, otherwise known as the "Philippine Librarianship Act," only graduates of the following degrees were allowed to take the Librarian Licensure Examination (LLE):

1. Bachelor of Library Science or Information Science.
2. Bachelor of Science in Library Science or Information Science.
3. Bachelor of Science in Education major/specialization in Library Science, or Bachelor of Arts major in Library Science.
4. Master of Library Science or Information Science.
5. Master of Arts in Library Science.

When Republic Act 9246, otherwise known as the "Philippine Librarianship Act of 2003" was passed, it further limited the library science courses that qualified graduates to take the Librarians Licensure Examination. Only graduates of the following courses were allowed to take the LLE.

• Bachelor of Library Science and Information Science.
• Master of Library and Information Science.

In view of the limitation in the degrees that qualify graduates for the licensure examination and to ensure quality graduates, the Professional Regulatory Board for Librarians (PRBFL) also developed the curricula for the two regulated degrees. The recommended curricula are now with the Commission on Higher Education (CHED) for review and approval. Upon the approval of the curricula and related provisions by CHED all library schools will have to comply with the course requirements and other requirements to be recognized as a course requiring licensure examination for their graduates. Within 5 years upon the effectivity of RA 9246, graduates of formerly approved courses were also still allowed to take the licensure examination. These courses were:

• Bachelor of Science in Education or Elementary Education or Bachelor or Arts with a major or specialization in Library Science.
• Master of Arts in Library Science or Information Science.
• Any master's degree with concentration in Library Science.

After the 5-year period, only graduates of the two regulated degrees, namely, Bachelor of Library and Information Science and Master of Library and Information Science will be allowed to take the Licensure Examination. Those with degrees other than the two but have formerly taken the licensure examination and failed to pass it will be allowed to take future licensure examinations by virtue of their having qualified to do so in the past.

Prior to RA 6966, government employees were required to take the Civil Service Examination given by the Bureau of Civil Service as a requirement for employment. On June 19, 1959, the Bureau of Civil Service was reorganized and renamed Civil Service Commission. The Commission conducted five examinations for librarians 1961, 1966, 1973, 1974, and 1981 for the positions of Supervising Librarian, Librarian, and Library Assistant. Those who passed these examinations were allowed to apply for licenses without examination under the provision of the "grandfather's clause" of the Republic Acts.

On June 22, 1973, the Professional Regulation Commission (PRC) was created by Presidential Decree No. 223. It was amended by Presidential Decree No. 657 on February 19, 1975 (Santos, 2003). On December 5, 2000 Presidential Decree No. 657 was repealed by Republic Act No. 8981, otherwise known as the "PRC Modernization Act of 2000 (Republic of the Philippines, 2000). Section 2 provides the policy framework for the Act.

Section 2. Statement of Policy. The state recognizes the important role of professionals in nation-building and, towards this end, promotes the sustained development of a reservoir of professionals whose competence has been determined by honest and credible licensure examinations and those standards of professional service and practice are internationally recognized and considered world-class brought about by regulatory measures, programs, and activities that foster growth and advancement.

In view of the PRC Modernization Act every professional regulatory law, has to be administered and enforced by the PRC (Santos, 2003). Thus for the library profession, the RA 6966 or the "Philippine Librarianship Act" had to be administered by the Professional Regulatory Board for Librarians (PRBFL) under the administrative control and supervision of the Professional Regulation Commission (PRC).

On February 19, 2004, Republic Act no. 9246 or the Philippine Librarianship Act of 2003 repealed RA 6966 (Republic of the Philippines, 2004). The new law is again administered by the PRBFL under the administrative control and supervision of the PRC. RA 9246 provides for the following statement of policy and objectives:

Article 1. Section 2. Statement of Policy. The State recognizes the essential role of librarianship as a profession in developing the intellectual capacity of the citizenry, thus making library service a regular component for national development. It shall, through honest, effective, and credible licensure examination and regulatory measures, undertake programs and activities that would promote and nurture the professional growth and well-being of competent, virtuous, productive, and well-rounded librarians whose standards of practice and service shall be characterized by excellence, quality, and geared towards world-class global competitiveness.

Article 1. Section 3. Objectives. This Act shall govern the following:

1. National examination for licensure, registration of librarians, issuance of Certificate of Registration and Professional Identification Card.
2. Supervision, control, and regulation of the practice of librarianship.
3. Integration of librarians under one national organization.
4. Development of professional competence of librarians.

The new law also defined the scope for the practice of Librarianship as follows:

Article 1 Section 5. Scope of the practice of librarianship. Librarianship shall deal with the performance of the librarian's functions, which shall include, but not be limited to the following:

1. Selection and acquisition of multimedia sources of information which would best respond to clientele's need for adequate, relevant, and timely information.

2. Cataloguing and classification of knowledge or sources of information into relevant organized collections and creation of local databases for speedy access, retrieval or delivery of information.
3. Development of computer-assisted/computer-backed information systems which would permit online and network services.
4. Establishment of library system and procedures; dissemination of information; rendering of information, reference and research assistance; archiving; and education of users.
5. Teaching, lecturing, and reviewing of library, archives, and information science subjects, including subjects given in the licensure examination.
6. Rendering of services involving technical knowledge/expertise in abstracting, indexing, cataloguing and classifying' or the preparation of bibliographies, subject authority lists, thesauri, and union catalogues/lists.
7. Preparation, evaluation or appraisal of plans, programs, and/or projects for the establishment, organization, development, and growth of libraries or information centers, and the determination of library requirements for space, buildings, structures, or facilities.
8. Provision of professional and consultancy services or advice on any aspect of librarianship.
9. Organization, conservation, preservation, and restoration of historical and cultural documents and other intellectual properties.

It also provided for the creation of the Professional Regulatory Board for Librarians (PRBFL) under the supervision and control of the Professional Regulatory Commission. The law vested the PRBFL with the following powers, functions, and duties:

1. To promulgate and administer rules and regulations necessary to carry out the provisions of this Act.
2. To administer oaths in connection with the administration of this Act.
3. To adopt an official seal of the Board.
4. To issue, suspend or revoke the Certificate of Registration and Professional Identification Card or grant or cancel a temporary/special permit.
5. To look into the conditions affecting the practice of librarianship, and when necessary, adopt such measures as may be deemed proper for the enhancement and maintenance of high ethical, moral, and professional standards of librarianship.
6. To adopt policies and set the standards for all types of libraries, librarians and the practice of librarianship.
7. To ensure, in coordination with the Commission on Higher Education (CHED), that all institutions offering library, archives and information science education comply with prescribed standards for curriculum, faculty and facilities for course offerings of library science, or library and information science.
8. To adopt and prescribe a Code of Ethics and a Code of Technical Standards for Librarians.
9. To hear and decide administrative cases involving violations of this Act, its Implementing Rules and Regulations or the Code of Ethics or the Code of

Technical Standards for Librarians; and for this purpose, issue subpoena ad testificandum and subpoena duces tecum to ensure the appearance of witnesses and the presentation of documents in connection therewith.

10. To prescribe guidelines in the Continuing Professional Education (CPE) in coordination with the accredited and integrated association for professional librarians.
11. To prepare, adopt, issue or amend the syllabi or items of specification of subjects for the librarian licensure examination consistent with the policies and standards set by the CHED.
12. To discharge other powers and duties as the Board may deem necessary for the practice of librarianship and the continued growth and development of librarians, libraries and library education in the Philippines.

The policies, resolutions, rules, and regulations issued or promulgated by the PRBFL or Board shall be subject to review and approval of the PRC. However, the Board's decision, resolution, or order rendered in administrative case shall be subject to review only if on appeal.

Another responsibility of the PRBFL not mentioned above is stated in Article III Section 14 of RA 9246.

Article III Section 14 states that:

Applications for registration, except those specifically allowed under *Section 19* of this Act, shall be required to undergo a written licensure examination to be given by the Board in such places and dates the Commission may designate subject to compliance with the requirements prescribed by the Commission.

Thus once every year, the Board prepares examination questions, conducts the examination, checks the answers of the examinees, and publishes the results.

The scope of the examination under RA 9246 includes:

1. Selection and acquisition of multimedia sources of information.
2. Cataloguing and classification.
3. Indexing and abstracting.
4. Reference, bibliography and information services.
5. Organization, management and development and maintenance of multimedia based library or information service, laws, trends and practices affecting the profession.
6. Information technology.

Just like in RA 6966 graduates and those practicing the profession may apply for exemption under RA 9246 if they qualify in accordance with the provisions of *Article III Section 19* which states:

Upon application and payment of the required fees, the Board shall issue a Certificate of Registration and Professional Identification Card to an applicant who, on the date of effectivity of Republic Act No. 6966, is:

(a) A practicing librarian who has completed at least a bachelor's degree, and a librarian or supervising librarian eligible.
(b) A practicing librarian who completed at least a bachelor's degree, 18 units in Library Science, 5 years experience in librarianship, and a first grade eligible or its equivalent.

(c) A practicing librarian who has completed a masters degree in Library Science or Library and Information Science, and a first grade eligible or its equivalent.

(d) A practicing librarian who has completed at least a bachelor's degree, 18 unit in Library Science, and 7 years experience in librarianship.

Those who qualified under this Section were given 3 years within which to apply for registration from the organization of the Board for Librarians.

RA 9246 also provides for foreign reciprocity thus,
Section 8. Foreign Reciprocity

A librarian from another country shall be admitted for licensure examination, be issued a Certificate of Registration and Professional Identification Card and be entitled to the rights and privileges appurtenant to this Act, if the country or state he/she is a citizen of or subject, permits Filipino librarians of said country or state to practice: Provided, That the requirements for Certificate of Registration in said country or state are substantially the same as the requirements under this Act: Provided, further, That the law of such country or state grants Filipino librarians the same privileges as the citizens/subjects of that country/state.

Other provisions further ensure the quality and behavior of practicing librarians. *Article II Section 22* provides for the non registration of any successful examinee or qualified applicant without examination if

he/she has been convicted by a court of competent jurisdiction of any criminal offense involving moral turpitude or has been found guilty of immoral and dishonorable conduct after investigation of the Board, or has been declared to be of unsound mind. The reason for the refusal shall be set forth in writing.

Section 23 gives the Board the

power after due notice and hearing, to revoke or suspend the Certificate of Registration or cancel a temporary or special permit of any librarian on any ground stated under Section 22 of this Act, or for any of the following: unprofessional or dishonorable conduct; practice of librarianship; fraud; deceit or falsification in obtaining a certificate of registration, professional identification card, or temporary/special permit; abetment of illegal practice by allowing illegal use of his/her certificate of registration, or professional identification card, or temporary/special permit; practice of profession during the period of suspension; or any violation of this Act, its Implementing Rules and Regulations, the Code of Ethics or the Code of Technical Standards for Librarians, or Board policies. The respondent may appeal the Board's decision, order/resolution to the Commission within 15 days from receipt thereof.

8.5   Code of Ethics for Registered Librarians

The Professional Regulatory Board for Librarians (PRBFL) approved a *Code of Ethics for Registered Librarians* through Resolution No. 2 dated August 14, 1992 (Santos, 2003). This was superseded by the Code of Ethics for Registered Librarians by Resolution No. 6, series of 2006 (Republic of the Philippines, 2006). The new Code of Ethics consists of a Preamble and four sections. The Preamble emphasizes the librarian's role in the development of knowledge and culture and the enrichment of people's lives and his/her responsibility to pursue the highest standards of behavior in relation to society, colleagues, and the profession. The first section specifies behavior towards the state, society and the public. The second section describes behavior towards the profession. The third section prescribes behavior towards suppliers, publishers, dealers, etc. The last section prescribes behavior towards clients and/or other users of their professional service. The Code of Ethics further specifies the disciplinary action that will be imposed after due process for violation of any provision in the Code which could be either revocation of his/her Certificate of Registration or suspension.

8.6   Curriculum Development

Article II Section 8 item g, of RA 9246 states one of the powers of the PRBFL, namely,

> to ensure, in coordination with the Commission on Higher Education (CHED), that all institutions offering library, archives and information science education comply with prescribed standards for curriculum, faculty and facilities for course offerings of library science, or library and information science.

To be able to comply with this provision, the PRBFL created a committee to develop standard curricula for the Bachelor of Library and Information Science (BLIS) and Master of Library and Information Science (MLIS) degrees. As of this date, the Commission on Higher Education has yet to approve the proposed curricula. In general however, the recommended BLIS degree will have general education courses, core courses, major courses and electives. Since the scope of the Librarians Licensure Examination (LLE) is prescribed by RA 9246, it is expected that the curricula for both the undergraduate and graduate levels will include (1) Selection and acquisition; (2) Cataloguing and classification; (3) Indexing and abstracting; (4) Reference, bibliography and information services; (5) Organization, management and development of library service, laws, trends and practices affecting the profession; and (6) Information and communications technology.

The inclusion of general education units required by the Commission on Higher Education (CHED) is designed to provide graduates not only with professional and personal competencies but also with well rounded knowledge in the Humanities, Sciences and Social Sciences. The professional and personal competencies are designed to produce graduates who are prepared to manage the needs of the twenty-first century library user. A thesis component which is also a CHED requirement aims to prepare the student for research after graduation while a practicum component will expose the student to the different facets of library work to prepare him for employment.

A study by David and Perez (2006), on the perception of licensed librarians about their academic preparation and job satisfaction, indicated a need to reexamine the curriculum to better prepare the graduates for the Librarians Licensure Examination (LLE). The practicing registered librarians also perceived that although they were academically prepared for entry into the workplace they found a need for continuous upgrade of their knowledge and skills to enable them to cope with the rapidly changing information environment. The respondents perceived a gap between theory and practice and so they fill in this gap by attending seminars and conferences. One of the weaknesses identified in the curriculum is in the area of management. They also expressed the need for an orientation program for new hires.

8.7 Continuing Education

Part of the duties and responsibilities of the Professional Regulatory Board for Librarians (PRBFL) is to prescribe guidelines in the continuing professional education programs in coordination with the Accredited Professional Organization (APO) for Librarians. In this regard a Continuing Profession Education Council (CPEC) was created by the PRBFL to oversee the continuing education programs being offered by various organizations. The CPE Council is composed of three members. One of the members come from the PRBFL, the second member is from a library school or the president of the Philippine Association of Teachers of Library and Information Science (PATLS) and the third member is the president of the Philippine Librarians Association, Inc. (PLAI). The CPE Council is now called the Continuing Professional Development Council (CPDC) under the provisions of the "Revised Guidelines on the Continuing Professional Development (CPD) Program for all Registered and Licensed Professionals" (PRC Resolution No. 2013-774).

Most of the continuing education programs for librarians are conducted by library associations. PLAI which is the Accredited Professional Organization for Librarians (APO) is the leader in this regard. PLAI sponsors a number of activities for librarians throughout the year. Its major event is the conduct of a national conference back to back with the general assembly. It also sponsors the "National Book Week" celebration in November of each year which culminates in a fellowship Lunch/dinner. Its major achievement however is the passage of RA 6966, the first Philippine Librarianship Act "regulating the practice of librarianship and prescribing the qualifications of librarians."

The PLAI has an executive vice president and vice presidents for each of the island groups Luzon, Visayas, and Mindanao. It also has eight regional councils. The Councils are situated in strategic regions of the Philippines so that they can reach out to the librarians in far flung regions of the country. The Philippine Librarians Association was originally founded as the Philippine Library Association (PLA) on October 22, 1923. It was renamed as the Philippine Library Association, Inc. (PLAI) on September 1, 1973 and again as the Philippine Librarians Association, Inc. (PLAI) on April 30, 1991 (Perez, 2004). A study by Villanueva (1997) on the perceived impact of PLAI conferences to the professional development of librarians indicate that members perceive PLAI sponsored conferences to have met their professional needs. Respondents recommend that PLAI continue sponsoring conferences and that it conduct more follow up studies regarding the impact of its conferences on the professional growth of its members.

Table 8.1 National library associations

Name of library association	Dates of founding
Association of Special Libraries of the Philippines	October 16, 1953
Public Libraries Association of the Philippines (PLAP)/ Now Philippine Public Libraries League (PPLL)	December 9, 1959/ June 30, 2004
Philippine Association of Teachers of Library Science (PATLS)/ Now Philippine Association of Teachers of Library and Information Science (PATLS)	February 28, 1964/ February 27, 2002
Agricultural Libraries Association of the Philippines (ALAP)/ Then Agriculture and Natural Resources Librarians Association of the Philippines/Now Agricultural Librarians Association of the Philippines (ALAP)	June 2, 1972/ October 18, 1996/ June 7, 2002
Philippine Association of Academic and Research Libraries/Now Philippine Association of Academic and Research Librarians (PAARL)	November 18, 1972/ January 30, 2004
Philippine Association of School Librarians (PASL)/Now Philippine Association of School Librarians, Inc. (PASLI)	June 1, 1977/ February 2, 1996
Philippine Group of Law Librarians (PGLL)	August 20, 1980
Philippine Theological Library Association (PTLA)	October 5, 1985
Medical and Health Library Association of the Philippines (MAHLAP)/ Now Medical and Health Librarians Association of the Philippines (MAHLAP)	May 1987
Court Librarians Association of the Philippines (CLAPHIL)	July 6, 2002

In addition to PLAI, other national library associations have been formed to respond to the needs of special groups. According to Perez (2004) there are seven types of library associations in the Philippines. These are (1) national, (2) regional, (3) provincial, (4) local, (5) library science alumni, (6) library-related, and (7) the student library associations. As of 2002, ten national library associations in addition to PLAI have been founded. The names and dates of founding and changes in names with corresponding dates of these national library associations are presented in Table 8.1.

All of these library associations are active in providing continuing education programs to their members. Unlike PLAI which only accepts licensed librarians as members, these other national library associations accept any library staff as member.

Library associations also cooperate with other international library associations in sponsoring international conferences. PLAI has cosponsored IFLA and CONSAL conferences in the Philippines. Other library associations also partner with international counterparts to hold conferences. In April, 2013, for example the Association of Special Libraries of the Philippines (ASLP) held an international conference (ICOASL2013) in cooperation with the Special Libraries Association (SLA) on the theme "Special Libraries towards Achieving Dynamic, Strategic, and Responsible Working Environment." Conferences such as this contribute a lot to the quality assurance of librarians in the Philippines.

Table 8.2 List of activities of selected National library associations in 2012

Name of library association	Activities in 2012
Agricultural Librarians Association of the Philippines (ALAP)	Seminar on "Evolving Practices in Professional Development and Sharing in the Digital Age"
Philippine Association of Teachers of Library and Information Science (PATLS)	Midyear National Forum on Library Education and Practice with the theme "Internationalization of Librarianship and Information Science: Challenges, Prospects and Promises"
Philippine Association of Academic and Research Librarians (PAARL)	• Lecture Forum on "RDA (Resource Description Access): A Guide to the Basics" • National Summer Conference on "Planning, Developing and Managing Digitization Projects and Researches for Libraries and Information Centers" • Public Forum on "Librarians Financial Management for Excellence" • Seminar Workshop on "Leadership 2013: the Challenge for the Information Professional" • Lecture Forum on "CANI (Continuous and Never-ending Improvement) on library and information services and the profession" • Lecture Forum on "Visions: the academic library in 2020 and PAARLNET" • Two orientation tours (One in-country and one in Malaysia
Philippine Association of School Librarians, Inc. (PASLI)	Seminar on "Digital School Library"
Philippine Group of Law Librarians (PGLL)	Seminar on "Popularizing the Law"
Medical and Health Librarians Association of the Philippines (MAHLAP)	Seminar on "Rethinking Services and Spaces: Rebranding Libraries for the Twenty-First Century" "Basic Photography Workshop for Librarians and Enthusiasts"

In addition to library associations, some libraries like the Rizal Library of the Ateneo de Manila University also sponsor international conferences which are attended by speakers and participants from many parts of the world. It has hosted six conferences since 2002. In 2012, it sponsored two conferences, one of which was in cooperation with the International Association of School libraries (IASL) with the theme "Directions for the Future of School Libraries." The second conference was on the theme "Libraries, Archives and Museums, Common Problems and Different Approaches."

Table 8.2 is a list of some seminars-workshops and conferences conducted by a selected group of national associations. PAARL is the most active among the national associations in conducting continuing professional education programs. It even conducts tours of libraries locally and abroad. It is also the biggest library association in the Philippines.

8.8 Exchange and Sharing of Expertise and Experience

To further assist librarians in enhancing their knowledge and skills, library associations and libraries invite experts to give lectures and seminars on trends and issues in Librarianship. At present libraries are concentrating on upgrading the knowledge and skill of librarians on Resource Description and Access (RDA). For example a series of seminars on RDA will be given by the PAARL in 2013 to reach out to the librarians outside the Metro Manila Area. (PAARL, 2013) The Library of the University of the Philippines—Diliman has also spearheaded the training on RDA through a cataloguing workshop held in 2012 (Tarlit, 2012). Benchmarking trips locally and abroad have also been spearheaded by individuals as well as library associations for purposes of making the librarians aware of trends and issues in libraries here and abroad. The "WOW Libraries" tour is designed "to provide librarians, library personnel and library enthusiasts the opportunity to see face to face the growth of outstanding academic libraries in the Philippines and observe the best library practices" (Aler, 2012). PAARL on the other hand has been sponsoring tours to other Asian libraries outside the Philippines. They have conducted tours in Malaysia in 2012, Hong Kong in 2011, and Singapore in 2010 (PAARL 2013).

In addition to conferences and seminars, libraries also accept trainees from library schools and libraries and hold training programs for librarians either within the library or outside upon the invitation of other groups.

8.9 Standards for Libraries

In addition to preparing standards for Schools offering Library Science programs, the Professional Regulatory Board for Librarians (PRBFL) also prepare standards for different kinds of libraries. The standards include qualifications required of librarians working in these libraries. The minimum standard is the Certificate of Registration issued by the PRBFL as proof that she/he is a licensed librarian. For school libraries, the head is also required to have graduate units in Library and Information Science. For academic libraries, the head librarian/director/university librarian is required to have a master's degree in Library and Information Science preferably with units at the PhD level. For special libraries subject specialization is also required in addition to the license. The licenses of librarians expire after 3 years and must be renewed upon presentation of evidence of continuing professional education/development programs attended within the 3-year period.

Aside from the PRBFL, other bodies also see to the quality of libraries and librarians in the country. Among these are:

- *Commission on Higher Education (CHED)*. The standards set by the CHED for the collection, facilities and services of academic libraries vary depending on the degree programs. However, with respect to the staffing, the minimum requirement is a registered or licensed librarian with a master's degree in Library and

Information Science with evidence of attendance in continuing education programs.

- *Library Associations.* The Philippine Association of Academic and Research Librarians (PAARL) is the leader in this endeavor. Although academic and research Libraries observe the standards set by PAARL, they must first meet the standards set by CHED and the Professional Regulatory Board for Librarians (PRBFL). The requirement set by PAARL for the head librarian is also a master's degree in Library and Information Science.
- *Accrediting Bodies.* The evaluation of the status of libraries is part of the accreditation programs of private universities, colleges or schools in the country. Government institutions may also opt to be accredited by these bodies.

The accreditation movement in the Philippines started in 1951. It aimed to enhance the quality of higher education through a system of standards. It is a quality management tool to ensure that institutions of higher learning have the minimum requirements for their programs including standards for human resources. The findings of the accrediting bodies are verified by CHED and the latter gives the final accreditation level for the school.

Three accrediting bodies were formed between 1951 and the seventies. These are the Philippine Accrediting Association of Schools, Colleges and Universities (PAASCU); the Philippine Association of Colleges and Universities—Commission on Accreditation (PACU-COA); and the Association of Christian Schools, Colleges and Universities—Accrediting Agency (ASCU-AA). Each of these accrediting bodies has their own instruments to measure performance of the educational institutions. In 1976, the Federation of Accrediting Agencies in the Philippines (FAAP) was established to serve as the coordinating body of the three agencies. Later a fourth accrediting body, the Accrediting Association of Chartered Colleges and Universities (AACCUP) was established to take care of accrediting state universities and colleges (Arcelo, 2003).

PAASCU was founded in 1957 by 11 private Catholic educational institutions. It operates on a scheme of program accreditation. It does not yet have an instrument to evaluate library schools, but it evaluates the human resources of the libraries of the educational institutions from basic education to tertiary level. It is a member of the International Network for Quality Assurance Agencies in Higher Education (INQAAHE). The minimum requirement for librarians in tertiary level educational institutions is a license or Certificate of Registration given by the Professional Regulation Commission (PRC). For the chief or head librarian a master's degree in Library and Information Science is also required.

ACSCU-AA began its accrediting work in 1971. Like PAASCU, it also focuses on accrediting programs instead of institutions. Its growth as an accrediting body is tied with the support of the Fund for Assistance to Private Education (FAPE) and the government. There is no exclusivity in accrediting programs in higher education institutions. ACSCU-AA may also accredit programs in institutions where some programs have already been accredited by PAASCU. It also requires a master's degree in Library and Information Science for the head librarian in academic institutions.

In 1967, PACU-COA published its *Handbook of Rules and Standards of Approval and Accreditation of Private Schools and Universities*. PACU-COA accrediting instruments are mainly in the fields of business, education, health sciences and ICT. The minimum requirement for librarians in the universities and schools accredited by PACU-COA is also a license and a master's degree in Library and Information Science.

AACUP is the accrediting agency for the member institutions of the Philippine Association of State Universities and Colleges. It began operations in 1989. The requirements for the head librarian for these institutions are also a license and a master's degree in Library and Information Science (Arcelo, 2003).

Since all these accrediting bodies require certificates of registration or licenses from the PRC, and a masters degree in Library and Information Science for the head librarian, practicing librarians are forced to pursue further studies to be promoted to higher levels.

FAAP is the coordinating body for the four accrediting agencies. It was launched in 1977 with only the first three accrediting bodies (PAASCU, ACSCU-AA, and PACU-COA) as members. They were later joined by AACUP. Although in 1979, it was officially recognized by the Ministry of Education, Culture, and Sports as the coordinating body for the accrediting institutions, it was only in 1984 that it was designated by the Department of Education as the official certifying agency for the four accrediting bodies. When the Commission on Higher Education (CHED) was created and in 1995 it encouraged the use of a "voluntary non-governmental accreditation system in aid of its regulatory function."

Accreditation as practiced in the Philippines is a voluntary mechanism for evaluating the status of programs in higher education institutions in several areas like faculty, instruction, administration and the library among others. To be accredited, one must have complied with the minimum standards set by CHED and the PRBFL since the accrediting bodies have a higher set of standards for some of the areas.

8.10 Regulatory Bodies

The Commission on Higher Education (CHED) is mandated to set and enforce minimum standards for program offerings of the higher educational institutions. They conduct accreditation activities but they also recognize the FAAP certified accreditation of the four accrediting bodies mentioned above. Under CHED accreditation, the qualifications of librarians and their continuing education activities are examined. The minimum requirements for librarians is the license and all are encouraged to obtain master's degrees in Library and Information Science or other degrees relevant to the course offerings of the institution if the undergraduate degree of the librarian is in Library and Information Science. Units if not candidacy or degree is also encouraged at the PhD level.

Prior to the passage of RA 8292 or the "Higher Education Modernization Act," state universities and colleges with the exemption of the University of the Philippines have to secure approval from the Civil Service Commission on the appointment of faculty and from the Department of Budget and Management for budgetary items (Republic of the Philippines, 1997). With the ratification of RA 8292, the state colleges and universities enjoy full academic freedom and fiscal autonomy as well as deregulated status per CHED Memorandum of January 2001 (Republic of the Philippines, 2001). Private institutions however, must secure recognition for their courses of study and should comply with the curriculum, and directives on curricular matters, physical facilities, library, etc.

Aside from the CHED, the Professional Regulation Commission (PRC) oversees the preparation of the graduates for the workforce. It administers the "licensure examination" to 46 professional fields including that for Library Science. The body that administers the examination for graduates of Library and Information Science is the Professional Regulatory Board for Librarians (PRBFL).

The first Librarians Licensure Examination was given in 1992. Since then until 2012 the PRBFL has conducted 21 licensure examinations and conferred more than 6000 licensed librarians. This number however has gotten smaller due to the demise of some of its members and transfer to other countries such as Singapore, Malaysia, the Middle East, Australia and the USA among others.

The average passing rate for 2007 to 2011 was 28 %. This increased in 2012 to 47 % (Filipino Librarian, 2010) and in 2013 to 45.80 %. The top performing schools in the November 2013 Licensure Examination as per Commission Resolution No. 2010-547 series of 2010 with 30 or more examinees and with at least 80 % passing percentage were the University of the Philippines (97.30 %) and the University of Santo Tomas (81.81 %) (Republic of the Philippines. Professional Regulation Commission, 2013). The schools with the top ten successful examinees were the University of Saint Louis in Tuguegarao (first place), University of the Philippines—Diliman, (second, fourth, sixth, seventh, and eighth places), Philippine Normal University (third place), Saint Mary's University (fifth place), Polytechnic University of the Philippines—Sta. Mesa (seventh place) University of San Jose Recoletos (ninth place), University of Santo Tomas and University of the East-Manila (tenth place). The data suggests that these schools should be visited and assessed to find out why some of their graduates are able to make it to the top ten.

There are also some schools whose graduates consistently fail to pass the examination. The data suggest that many of these schools are not delivering quality Library and Information Science education. These are also mostly schools with very low enrollments hence the questions: (1) do the LIS programs in these schools deserve to remain open? (2) is there really a need for the more than 100 schools offering LIS in the country?

In 2012, *The Roadmap to Quality Professional Regulation* as the short and medium-term strategic plan of the Professional Regulation Commission (PRC) was approved. As regulatory bodies, CHED and PRC issued a circular requiring state and local universities and colleges to secure authority from CHED to operate programs requiring Board examinations and for PRC to admit only applicants for licensure examinations effective January 2011 from educational institutions with authority to offer such programs. This could have had an effect on the increase in passing rate in the LLE 2012 and 2013 examinations.

8.11 Awards and Recognition

Awards and recognition inspire librarians to do better and deliver quality service. The Professional Regulation Commission (PRC) recognizes librarians who have done wonderful work in the field. The Professional Regulatory Board for Librarians (PRBFL) recommends to the PRC, nominees for Outstanding Librarian of the Year upon the recommendation of the Philippine Librarians Association, Inc. (PLAI), the accredited professional organization for librarians. From 1992 to 2012, the Professional Commission has recognized 19 outstanding librarians. The Outstanding Professional of the Year Award is the "highest award given by the PRC to a registered librarian for having demonstrated exceptional professional competence and integrity in the practice of one's profession as recommended by one's peers, and contributed significantly to the advancement of the profession and to the effective discharge of the profession's social responsibility through meaningful participation in socio-civic related activities" (PAARL 2012). The conferment of the award was discontinued from 1986 to 1991 (Republic of the Philippines, 2011) but has since been revived.

Other recognitions are given by the library associations. The PAARL annual awards are the most prestigious of these awards for academic and research libraries and librarians. These awards are given; "a) to foster the professional growth of academic/research librarians; and b) to give recognition for special achievements." Among these awards are:

- The Academic or Research Librarian of the Year Award which is given to an individual librarian who has made an outstanding national contribution to academic or research librarianship and to library development.
- The Outstanding Library of the Year Award which is given to a research or academic library in recognition of its outstanding national contribution to academic or research librarianship and library development in the Philippines.
- The Lifetime Achievement Award which is given to an academic/research librarian as a hallmark of professional excellence. It is the highest recognition given by PAARL to an academic/research librarian.
- The Outstanding Library Program of the Year Award which is given to an academic/research in recognition of its outstanding library program that contributes to academic librarianship and library development in the Philippines (PAARL, 2009).

8.12 Comparison with Other Countries of the Region

Compared with other countries in the region, the Philippines is perhaps the only country that has a unique system of ensuring the quality of library education and practice by virtue of the Republic Act 9246 otherwise known as the "Philippine Librarianship Act of 2003." In addition the law is implemented through regulatory bodies that provide for a licensure examination, and a standard curriculum.

Further monitoring is carried out through accrediting bodies that require minimum qualifications for the librarians in academic, school, public, and special libraries. Library Associations further ensure quality through continuing education programs. Perhaps the only aspect lacking in ensuring quality in library education and practice in the country is the absence of a doctoral degree in Library and Information Science. At this writing however, a number of Filipino librarians have returned with doctoral degrees obtained from other countries. It is foreseen that a doctoral degree in Library and Information Science will soon be offered locally.

8.13   Impact of the Law on Librarians and the Practice of Librarianship in the Philippines

The regulation of the practice of librarianship in the Philippines has had positive impact on the image of the librarian and in the practice of librarianship. Several studies support this observation.

Before the passage of Republic Act 6966 and the new Philippine Librarianship Act (RA 9246) any individual with or without a college degree can be appointed to the post of librarian. Such individuals were given salaries equivalent to that of clerks and other nonacademic personnel. With the regulation of the practice of librarianship the librarians enjoyed being classified with the academic personnel and given salaries and stature equivalent to that of the faculty. With the new law librarians were recognized as professionals and were given appointments as teaching faculty or research faculty (Dizon, 2003). A follow—up study of graduates of Library and information science from 1996 to 2005 by Bacharo, (2007) indicate a very good profile of employment for librarians.

At the Ateneo de Manila University Loyola Schools, librarians are classified as professionals and enjoy benefits and salaries similar to that of the faculty. Alongside this recognition however, librarians also have to pursue higher degrees in order to be promoted in rank. The license to practice is an entry level requirement but a master's degree is a requirement for promotion (Ateneo de Manila University, 2009). A similar ranking and promotion scheme for librarians of the University of the Philippines is also observed.

In addition, attendance in continuing professional education/Development (CPE/CPD) programs is necessary before one can apply for the renewal of the license to ensure that the librarian is aware of developments in her profession. In the beginning the librarians did not like attending CPE programs because of the expense involved but eventually, they saw its importance and value (Beltran, 2007; Tabiendo, 2003). One good development in this regard is the recognition by employers of the need for librarians to be updated with developments hence many employers provide financial assistance to their librarians so that they can attend such CPE/CPD programs.

Another positive impact of the license and CPE/CPD requirements is the increasing number of librarians conducting research, presenting papers in international conferences and publishing papers. In the past librarians were only involved in running their respective libraries and were not doing research. A search of Scopus will yield several hits for Filipino librarians who have published articles in international peer reviewed journals. Related to this is the recognition given by employers to librarians who conduct research, present and publish papers which is similar to that of faculty. This practice is observed in at least three universities, namely, the Ateneo de Manila University, De La Salle University and the University of the Philippines.

One very positive impact is the monitoring of quality assurance in library schools by the Board for Librarians. The national passing average increased from 27 % in 2010 and 28 % in 2011 to 47 % in the 2012 and 45.80 % in 2013 licensure examination due to this monitoring system (Professional Regulation Commission). Likewise the number of non-performing schools also decreased from 60 in 2010 and 45 in 2011 to 32 in 2012 with a slight increase to 36 in 2013. (Republic of the Philippines, Commission on Higher Education. Official Gazette). In monitoring the performance of schools, representatives of the CHED and the PRBFL visit the schools to assess the performance of the teachers of Library and Information Science and ensure that graduates are ready to be admitted into the workforce.

Hand in hand with monitoring the performance of the schools, the Board also provides standards for the libraries from basic education to tertiary level as well as for special and public libraries.

Overall, the major impact is on the image of the librarian and his/her position in the academe and society (Wee, 2000; Sobrevinas, 2007; Visperas, 2002). The improvement in the self esteem of the librarians is remarkable. They now work in partnership with the faculty and researchers whether they are working in academic, special or public libraries. The Philippines still has a long way to go to fill its needed quota of librarians for its libraries but the number of applicants for admission in the top performing schools in the licensure examination is increasing. The law has professionalized librarianship and the Filipino librarian.

8.14   Conclusion

The quality assurance measures for library education and librarianship in the Philippines are in place with the enactment of the Republic Act 6966 and Republic Act 9246 and the creation of the Professional Regulatory Commission (PRC), the Professional Regulatory Board for Librarians (PRBFL) and the Commission on Higher Education (CHED). The provisions of the law and its implementation coupled with the regulatory functions of the PRC and CHED provide guidelines to ensure quality in the library profession. With the help of library associations continuing education activities are made available to the professionals. However, librarians who would like to pursue doctoral studies in Library and Information Science (LIS) can only do so if they get fellowships in universities abroad. At present there are already a number of librarians with PhD in LIS in the Philippines. A few more will be returning soon. It is foreseen that within a few years a doctoral program will be offered locally.

References

Aler, O. (2012). *Wow Libraries! Tour of Outstanding Libraries goes to SSC*. http://www.ssc.edu. ph/sscweb/news2012/2012_0928_WOW_LIBRARIES_Tour_SSC.html

Arcelo, A. (2003). *In pursuit of continuing quality in higher education through accreditation: The Philippine experience*. Paris: UNESCO International Institute for Educational Planning.

Ateneo de Manila University Loyola Schools (ed.), (2009). *Professionals manual*. Quezon City: Ateneo de Manila University.

Bacharo, M. B. (2007). *Employment profile of the Bachelor of Library and Information Science graduates from 1996-2005: a follow-up study*. (Thesis) Quezon City: University of the Philippines, School of Library and Information Science.

Beltran, C. R. C. (2007). *Continuing professional education of University of the Philippines Diliman Librarians: revisiting of recommendations*. (Thesis) Quezon City: University of the Philippines, School of Library and Information Science.

Chambers, E. (1960). *Library Education in the Philippines: An Informal Survey*. Urbana, Illinois: University of Illinois Library School.

Damaso, C. (1966). Library Education in the Philippines. *Journal of Education for Librarianship*, *6* (4) Spring

David, L., & Perez, D. (2006). An assessment of the perception of licensed librarians about their academic preparation and satisfaction in their job as librarians. In *Proceedings of the Asia-Pacific Conference on Library and Information Education & Practice 2006 (A-LIEP 2006) Preparing Information Professional for Leadership in the New Age*. Singapore, (pp. 416–422) 3–6 April 2006.

Dizon, M. C. (2003). *Employment status and benefits of teaching and non-teaching academic librarians and regular faculty*. (Thesis) Quezon City: University of the Philippines, School of Library and Information Science.

Filipino Librarian. Librarian's Licensure Examination. (2010). *Results*. Retrieved at http:// filipinolibrarian.blogspot.com/search?updated-min=2010-01-01T00:00:00%2B08:00&updated-max=2011-01-01T00:00:00%2B08:00&max-results=13

PAARL. (2009). *PAARL Awards*. https://sites.google.com/site/paarlonlineorg/home/home/awards/paarl-awards-1

PAARL's National Summer Conference on Resource Description and Access. (2013). *Davao City*. 22–24 April 2013. https://sites.google.com/site/paarlonlineorg/

Perez, D. (2004). *Philippine Libraries and Librarianship, 1900–2000: a Historical Perspective*. (Unpublished Thesis) Quezon City: University of the Philippines-Diliman, Institute of Library Science.

Philippine Islands. (1918). Public Laws Enacted by the Philippine Legislature during the Period March 17, 1917 to May 29, 1918 Comprising Acts nos. 2711 to 2780, (Volume 13): Including a Numerical List of Acts; a General Lists of Repealed and Amended Acts; a List of Codes, General Orders, etc., Amended; Joint and Concurrent Resolutions of the Philippine Legislature. Manila: Bureau of Printing.

Republic of the Philippines. (2011). PRC Official Website. *History* http://www.prc.gov.ph/about/default.aspx

Republic of the Philippines. Commission on Higher Education. (2001). *Memorandum Order 03 series of 2001*. Revised Implementing Rules and Regulations (IRR) for Republic Act 8292, otherwise known as an "Act Providing for the Uniform Composition and Powers of the Governing Boards, the Manner of Appointment and Term of Office of the President of Chartered State Universities and Colleges, and for other Purposes." Jan 17, 2001.

Republic of the Philippines. Congress of the Philippines. (2000). *Republic Act 8981*. An act Modernizing the Professional Regulation Commission, Repealing for the Purpose Presidential Decree Numbered Two Hundred and Twenty-three Entitled "Creating the Professional Regulation Commission and Prescribing its Powers and Functions and for other Purposes.

Republic of the Philippines. Professional Regulation Commission. (2006). Board for Libraries Resolution no. 6. Series of 2006."Code of Ethics for Librarians."

Republic of the Philippines. Congress of the Philippines. (1990). *Republic Act 6966*. An Act Regulating the Practice of Librarianship and Prescribing the Qualifications of Librarians. Eighth Congress, September 19, 1990.

Republic of the Philippines. Congress of the Philippines. (1997). *Republic Act 8292*. An Act Providing for the Uniform Composition and Powers of the Governing Boards, the Manner of Appointment and Term of Office of the President of Chartered State Universities and Colleges, and for other Purposes. Tenth Congress, June 6, 1997.

Republic of the Philippines. Congress of the Philippines. (2004). Republic Act No. 9246 "An Act Modernizing the Practice of Librarianship in the Philippines thereby Repealing Republic Act 6966, Entitled "An Act Regulating the Practice of Librarianship and Prescribing the Qualifications of Librarians," Appropriating Funds Therefor and For OtherPpurposes. Twelfth Congress Third Regular Session, February 19, 2004

Sanchez, C. (1966). *Education for Librarianship in the Philippines: its History, Development and Status*. (Unpublished Thesis) Manila: Centro Escolar University.

Santos, A. (2003). *The Professionalization of Librarians in the Philippines: the Role of Library Associations*. World Library and Information Congress: 69th IFLA General Conference and Council, Berlin, 1–9 August 2003.

Sobrevinas, M. G. C. (2007). *Perceptions of selected library users on the library profession*. (Thesis) Quezon City: University of the Philippines, School of Library and Information Science.

Tabiendo, G. B. (2003). *The correlation between the demographic profile and continuing professional education of academic librarians in Metro Manila*. (Thesis) Quezon City: University of the Philippines, School of Library and Information Science.

Tarlit, R. (2012). Background and Overview of RDA. *RDA Workshop*. Quezon City: University of the Philippines-Diliman: University Library, May 7, 2012. http://www.mainlib.upd.edu.ph/?q=cataloging-workshop

University of the Philippines. Board of Regents. (1922). Minutes of the 223rd Meeting of the Board of Regents.

Villanueva, A. (1997). *Impact of PLAI Conferences, etc, to the Professional Development of Librarians and Information professional, 1991–1995*. (Unpublished Thesis) Quezon City: University of the Philippines, Institute of Library Science.

Visperas, R. S. (2002). *Faculty perceptions of academic librarians at the University of the Philippines Diliman*. (Thesis) Quezon City: University of the Philippines, School of Library and Information Science.

Wee, M. M. L. (2000). *Job satisfaction among licensed librarians along Herzberg's Hygiene-Motivation theory in the Philippine setting*. (Thesis). Quezon City: University of the Philippines, School of Library and Information Science.

Chapter 9
LIS Education and Quality Assurance System in Asia-Pacific Region Thailand: Recent Trends and Issues

Sujin Butdisuwan

9.1 Introduction

The role of an educational institution is to develop independence of thought, logical thinking and finally become a contributing member of the society. The twenty-first century focuses on building knowledge societies and the higher education is expected to play a major role in the knowledge driven global economy. As knowledge can be created, absorbed and applied by the educated minds, academic institutions will have to be effective through the activities of discovery, shaping, achieving, transmitting, and applying knowledge. Globalization has opened the gates of higher education to become an international service. There is growing concern all over the world about quality, standards, and recognition. Because of this trend, it is significant to examine how best practices benchmarks have to be evolved for ascertaining and assuring quality at different levels of higher education in different countries.

To provide quality education to large numbers at affordable costs is the primary concern of developing countries to compete at global level and Thailand is not an exception. Quality education should be socially relevant and personally indispensable to the individual. Hence, quality and excellence need to be the vision of every higher education institution. It is a great challenge faced by most of the higher education institutions in the developing world. The overall aim of education is to develop lifelong learners; best practice principles should enable this transformation to take place in any subject area.

This millennium brought in new challenges and opportunities to the library and information science education apart from other disciplines in Thailand. Training for library and information professionals is essential for effective functioning of the libraries and information centers. Education and training should be based upon

S. Butdisuwan (✉)
Mahasarakham University, Maha Sarakham, Thailand
e-mail: sujin.b@msu.ac.th

M. Miwa and S. Miyahara (eds.), *Quality Assurance in LIS Education:*
An International and Comparative Study, DOI 10.1007/978-1-4614-6495-2_9,
© Springer Science+Business Media New York 2015

the demands of the profession. But, since the demands of the profession are always changing due to various factors, such as, growth of literature, complexity of subjects, change in the forms of documents, impact of ICT, etc, the training methods must adapt to such changes. LIS Education, as a professional course needs revamping to address the present and future developments in technology as a tool in identifying, organizing, storing, providing access facilities to the end user. It is necessary to identify the best practices in LIS Education to bring in uniformity in the teaching and learning process in the country. This write-up tries to capture the existing situation of quality assurance systems, guidelines, and standards to assess the quality and also to promote the concept of best practices in the institutions of higher learning in general and more importantly in Library and Information Science Education in Thailand.

9.2   Concept of Quality

One of the major problems plaguing the field of assessing quality is the inconsistent use of the term. Quality in LIS is a value judgment, differently interpreted by various stakeholders, such as governments, employers, students, administrators, and LIS teachers. Because quality is a very subjective concept, it is very important to identify the accrediting body in order to understand the procedures and purposes of the evaluation as well as to establish the authority and validity of the evaluation.

Quality definition focuses on excellence on consistency as determined by the stakeholders, who have an interest and on the accountability in terms of the efficiency and productivity of the evaluation process. Quality assurance is defined as a planned and systematic review process of an institution or program to determine that acceptable standards of education, scholarship, and infrastructure are being maintained and enhanced (CHEA, 2003). Quality Audit on the other hand is a test of an institution's quality assurance and control system through a self-evaluation and external review of its programs, staff, and infrastructure.

All the LIS guidelines are fairly open and flexible enough to offer space for different approaches (Khoo, Majid, & Sattar Chaudry, 2003). LIS guidelines cover the following areas:

The context of the program, the institutional support, the relationship with the parent institutions;

- Mission, goals, and objectives.
- Curriculum.
- Faculty and staff.
- Students and policy and procedures.
- Administration and financial support.
- Instructional resources and facilities.
- Regular review of the program.
- Employment and labor market.

Another approach to quality assurance in LIS is the application of industrial standards such as ISO 9000 series, and management systems such as TQM (Total Quality Management). The ISO 9000 series intends to stimulate trade by providing assurance of an organization's ability to meet specifications and perform the negotiated standards. TQM combines quality control, quality assurance, and quality improvement and goes beyond traditional customer satisfaction by addressing the needs of internal customers (as students, parents, and employers), suppliers, and other stakeholders.

9.3   Library and Information Science Education in Thailand: A Brief Note

Library education in Thailand was first introduced at Chulalongkorn University in 1951 under the support of the Fulbright Foundation. At the beginning, it was just a training program, conducted by five American professors who offered a certificate in Library Science. In 1955, the Department of Library Science was established at the Faculty of Arts, Chulalongkorn University to offer a program for a diploma in library science. At present, the universities both private and public are offering programs at a bachelor degree, master degree, and doctoral level (Butdisuwan & Gorman, 2002; Premsmit, 1999; Ruksasuk, 1999). Over a period of time, several institutions including public, private, and Rajabhat Universities started different LIS programs at undergraduate and postgraduate levels. LIS education programs have different names in different universities, for example, Library Science, Library and Information Science, Information Science, Information Studies, Information Management, and Information Technology and Management. However, all these programs fall under undergraduate and postgraduate courses. At present most universities offer undergraduate programs; 13 universities offer postgraduate programs and only Khon Kaen University and STOU are offering doctoral program in LIS in Thailand. In addition to these, two universities offer LIS education through distance mode, namely, Ramkhamhaeng University and Sukhothai Thammathirat Open University (STOU) and both located in Bangkok (Saccanand, 1999). Most of the undergraduate programs are offered in the faculties on various names such as Faculty of Humanities, Faculty of Humanities and Social Sciences, Faculty of Arts, and Faculty of Liberal Arts as full-time, part-time, or both. These programs are developed keeping in view of the requirements of library professionals at the entry level. The undergraduate program in LIS is for 4 year duration by most of the departments but some departments also offer 2-year duration program according to the standards set by Commission on Higher Education (former the Ministry of University Affairs). LIS graduate studies in Thailand include master degree program, advanced graduate diploma program and a doctoral program. Objectives of these programs are to prepare high level professionals to work in Libraries using ever changing technologies and also teach in LIS departments. Master degree program is offered as full-time or part-time in most of the universities. Rapid expansion in LIS education and changes in the economic and social climate in which this education is provided have raised numerous problems. The Thai LIS departments share common set of problems for which solutions must be sought cooperatively. These include staff capabilities, course content, student selection, and employment opportunities for graduates (Butdisuwan & Gorman, 2002).

There have been efforts in the direction to improve staff capabilities by sending them to other countries for higher qualifications (doctoral program) and training to strengthen the departments when they came back. Course content is a serious concern and it needs innovations, application of new technologies, addressing the contemporary issues using ICT etc. This may change gradually in many universities as and when the teachers comeback with higher qualifications and training etc. Apart from sending the faculty members to acquire higher qualification, the institutions can create facilities for in-service training programs specially designed to train the trainers. Selection of students and employment opportunities largely depend on the credibility of the courses that are taught based on the market requirements. Thai library professionals should also learn English to look at global opportunities. LIS education should focus on specialized areas to train the graduates to handle specialized works to work in various types of libraries in Thailand and also outside Thailand apart from improving communication skills and soft skills. It is equally important to learn from each other's experiences within or outside of Thailand. This is possible, if quality assurance systems are in place and a database of best practices that are implemented or adopted by various departments and added value to the course, are created and constantly revised as the changes that take place in library and information science education in Thailand. Library and information science education in Thailand is entering in to its 62nd anniversary. The number of institutions that offered the programs has continued to increase. Many institutions are in the process of curriculum review and revision as well as new program initiation. Still some constraints are reported in the following areas: the graying of faculty numbers, lack of extensive expertise in information technology, heavy teaching load, insufficient budget for instructional equipment and educational media development. Furthermore, library school personnel are also now concerned with recruiting high-caliber students into the programs (Butdisuwan, 2000).

9.3.1 Quality Assurance System in Thailand

In Thailand the quality assurance in higher education was first thought by the Ministry of University Affairs (MUA) during 1996. The MUA (then overseeing 23 public and 53 private higher education institutions) has announced its policy to encourage all public and private universities to establish and maintain quality assurance in teaching and research.

The introduction of the National Education Act in 1999 states that quality assurance in educational systems comprises of internal quality assurance (IQA) and external quality assurance (EQA). In order to maintain the IQA, each academic

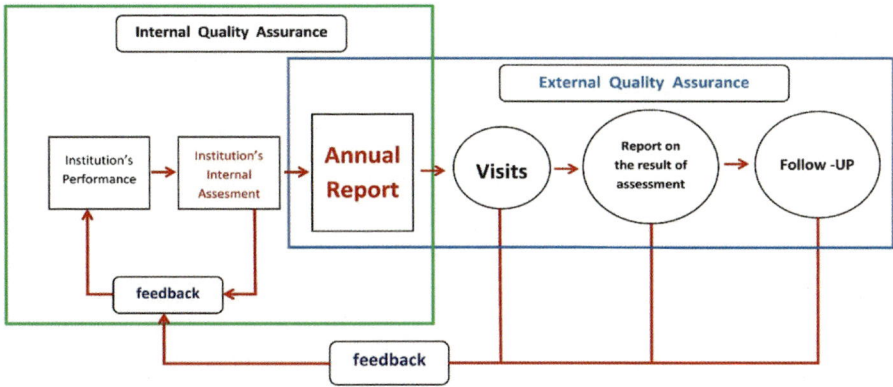

Fig. 9.1 Two processes to maintain quality assurance

institution and its governing organization have to oversee the internal mechanisms and continue to remain as a part of the continuing management system. On the other hand the Office of the National Education Standards and Quality Assessment (Public Organization) (ONESQA) is responsible for ensuring the external assessment of institutions at all levels.

The quality assurance is to be maintained by following the two processes, namely, Internal Quality Assurance (IQA) and External Quality Assurance (EQA), as indicated in Fig. 9.1. The IQA is the responsibility of respective institution which has to make a self-assessment as per the guidelines given that the Higher Education Commission analyzes the institution's performance based on the guidelines and reports in the annual reports every year. On the other hand, the EQA is done by the outside experts by visiting the respective institution once in 5 years, inspect the infrastructure facilities and examine the annual reports and IQA reports and prepare a detailed report on the external quality assessment and submit the evaluation report. Based on the evaluation report and the feedback, the institution will have to perform and strive for the maintenance of the quality assurance.

9.3.2 *Quality Assurance Framework by Office for National Education Standards and Quality Assessment (ONESQA)*

The Office of the National Education Standards and Quality Assessment (ONESQA) was established in 2001 as a public autonomous body which is responsible for ensuring the external quality assessment of all educational institutions. At the higher education level, ONESQA provides 28 indicators and to review the institutional performance in 8 categories such as:

1. Quality of Graduates (4 indicators).
2. Learning Process (4 indicators).

3. Learning Support Resources (5 indicators).
4. Research and Innovation (4 indicators).
5. Academic Services (2 indicators).
6. Preservation of Art and Culture (2 indicators).
7. Administration and Management (5 indicators).
8. Internal Quality Assurance System (2 indicators).

Some of the guiding principles set forth by ONESQA to ensure external assessment are as follows:

- Ensure that higher education is developed to the standards of international levels.
- Review and confirm existing system of the institute, assessing quality of each and every main function while keeping in mind their academic freedom, uniqueness, values, principles, missions, and goals.
- Operate under the objectives, principles, and directions set forth in the National Education Act.
- Assure and support the implementation of internal quality assurance system within the institution.
- Uphold the quality of academic standards in higher education institutes
- Employ amicable assessment procedures without lessening the integrity of transparency and accountability.

Awareness, attempt, and achievement are the three dimensions focused in the assessment process. Every institution has to submit relevant and required data along with the self-review report to the ONESQA before being subject to external assessment visit. Added to this, certain documents and reports on internal quality assurance system may be submitted to supplement overall understanding of the nature of institution prior to on-site visit. A team of external reviewers comprising of academicians in the related areas and disciplines make visits to the campuses according to a preplanned schedule. Based on the visit and examination, an evaluation report together with findings and recommendations will be sent back to the institution.

9.3.3   Quality Assurance Framework by the Office of Commission of Higher Education (CHE)

Quality control, quality audit, and quality assessment are the prime concepts of quality assurance guidelines as proposed by MUA and presented in Fig. 9.2. The framework consists of nine components such as:

1. Philosophies, Commitment, and Objectives.
2. Teaching and Learning.
3. Student Development Activities.
4. Research.
5. Academic Services.
6. Preservation of Art and Culture.

Fig. 9.2 Nine components of quality audit and quality assessment guidelines proposed by MUA

7. Administration and Management.
8. Internal Quality Assurance System and Mechanisms.
9. Finance and Budgeting.

The above stated nine components are presented and explained in Table 9.1.

9.3.4 Thai Qualifications Framework (TQF)

The Qualifications Framework for Thailand's higher education system is designed to support implementation of the educational guidelines set out in the National Education Act, to ensure consistency in both standards and award titles for higher education qualifications, and to make clear the equivalence of academic awards with those granted by higher education institutions in other parts of the world. The Framework will help to provide appropriate points of comparison in academic standards for institutions in their planning and internal quality assurance processes, for evaluators involved in external reviews, and for employers, in understanding the skills and capabilities of graduates they may employ.

Table 9.1 The nine components of quality indicators and assessing guidelines for IQA

Component	Example of indicator	Example of focusing area
1. Philosophy, mission, objective, and implementation plan	• Determined philosophy, missions, objectives, implementation plan, and standard accomplishment indicators • Percentage of the standard accomplishment based on the indicators assigned	• Indicated philosophy or aspiration • Correspondence of philosophy, mission, objective, and strategy with the main institutions' works and activities, as well as other standards
2. Teaching and learning provision (academic affairs)	• Student is the center of teaching and learning • No. of full-time students to full-time professors/instructors • Percentage of professors/instructors with BA/MA/PhD degrees	• Specifying the upcoming courses and closing down courses • Students as a center of learning • Follow-up on the students' career accomplishment
3. Student development activities	• Determining and evaluating on the activities and facilities provided for students, e.g., IT, student housings and dormitories, environment, canteens, information on scholarships and job opportunities for students, etc.	• Surveying on the need of students • Providing consultants/counselors for students • Organizing career advancement campaign (provide training) for students or alumni • Media center for students/alumni
4. Research	• Acquired system and mechanism development that support students in conducting researches • Internal and external funding on research development and creative works for professors • Percentage of research and creative works published, publicized, and/or utilized at national and international levels in the proportion to the total number of full-time instructors	• Mechanism on human resource management for conducting research • Research administration and supplements • Supporting mechanism on conduction, promoting, publishing research
5. Academic service to community	• Developed system and mechanism for academic service to community to the standard requirements • Percentage of full-time professors/instructor participating in academic service to community; percentage of communities satisfactory on academic services	• Working/managing according to policy and planning • Evaluation on the administration in academic service • Integration between academic service, learning, and preservation of arts and culture

(continued)

Table 9.1 (continued)

Component	Example of indicator	Example of focusing area
6. Preservation of arts and cultures	• Acquired system and mechanism development for preservation of arts and cultures • Produced products that aim to develop knowledge-based and to set standard on arts and cultures (only) institutions that specialized in arts and cultures)	• Policy and planning development • Promoting development • Supporting development • Setting standard development
7. Administration and management	• Leadership quality of institution administrators • Institution development for transformation into a learning organization • Percentage of professors, who are nationally/internationally awarded in academic areas • Effectiveness system for management; teaching; learning and research activities	• Participation on benchmarking vision, strategy, and policy of institution administration • Managing process of institutions • Follow-up on the main areas of administrative system • Seeking for leadership quality of the university council/executive/committee members of the institutions
8. QA system and mechanism	• System and mechanism for internal quality assurance as part of the institutions' governance • System and mechanism for knowledge and skill on quality assurance to students • Level of accomplishment on internal quality assurance	• Quality assurance system • Mechanism on quality assurance • Standard, indicators, and quality guideline • Database system for quality assurance
9. Finance and budgeting	• System and mechanism for allocating, analyzing expenses and auditing efficiency • Internal and external sharing of resources (level)	• Internal resources management policy • Database system on resources management • Follow-up and evaluation

The Thai Qualification Framework for Higher Education (TQF:HED) was launched in 2008. According to the National Education Act 2542 (B.E), the TQF: HED is a mechanism for higher education quality assurance and is intended to develop of the quality students in the higher education system.

The implementation of the TQF poses substantial challenges to learning and teaching. These include endemic difficulties in integrating theory and practice, and the shift of focus of activity and effort to documentation rather than the quality of teaching and learning itself. Other challenges relate to recent and significant changes in the nature of Thai higher education which have had considerable impact on the design of teaching strategies, learning activities, and assessment.

The TQF should include a process that supports flexibility in learning, teaching, and assessment of students from various backgrounds and disciplines. The key mechanism that will help academics in higher education system to be successful in implementing the TQF is teaching strategies and an evaluation system that articulates directly with the qualification framework.

Learning outcomes are statements of the attributes and capabilities that a student should have achieved on successful completion of the learning session or topic. They provide a reference point for assessing student progress and designing assessment strategies and methods. Learning outcomes are helpful benchmarks for the standards educators will apply when they measuring students' achievements using various assessments instruments and processes.

Drawing from evidence, the quality of graduates can be ensured by means of the implementation of sound teaching strategies tailored for students of diverse backgrounds. In addition, assessment tasks offer important opportunity to enhance students' learning experience. The TQF is aptly constructed around these proved notions and, as such, has strong potential to lead to full integration of learning experiences and quality. It is vital that the framework is translated into ground-level action and, therefore, that educators are intimately familiar with the TQF and how to apply it to practice.

9.3.5 Quality Assurance Practice in LIS Education

Internal quality assurance and accreditation of higher education through an external agency will make the institutions to compete with global level institutions. To achieve academic excellence, it has been envisaged that standards and norms for LIS education should be set by an external agency and thereafter adherence to them may be made mandatory. Accreditation is a set of processes whereby an outside agency evaluates and examines the LIS courses according to a set of predetermined norms and standards. The professional accreditation of LIS education in universities has long been practiced in the UK, USA and Australia (Enser & Wood, 1999).

9.3.6 Guidelines and Indicators (6) for Internal Quality Assurance

1. Covering all the components of ministerial regulations for the system, criteria, and methods for internal Quality Assurance (2003).
2. Reflecting the aims of Second National Education Act B.E. 2545 (2002), the indicators for External Quality Assurance of ONESQA and the Office of the Office of the Public Sector Development Commission (OPDC).
3. Assessing to all dimensions of Quality Assurance System (Input, Process, and Outcome).

4. Balancing four main management perspectives (Students and Interested persons, internal procedures, financial, and human resources, learning and innovation).
5. Minimizing number of indicators and guidelines so that institution can develop their own to match with the suitability.
6. Separating indicators and guideline to fit with two different types of institutions (For all institutions and for particular areas of expertise institutions).

It has now become imperative to establish an external agency at the national level which can undertake the work of accreditation of LIS courses in Thailand. The Council for Higher Education can take the lead in developing a mechanism for introducing a method of accreditation of LIS schools and courses. The Government should pass an Act that should make provision for establishing a statutory Library and Information Science Council in Thailand and it should function as a central agency to provide guidelines and standards in LIS education and practices. It would be responsible to lay down parameters for starting new LIS schools, continuation of existing LIS schools, recognition and equivalence of different levels of LIS degrees for the purpose of employment or higher studies, and promotion of LIS education in Thailand.

9.4 Quality Assurance Issues

9.4.1 Criteria for Assessment

The derivation of criteria can be of two types - by reviewing process and facility of education in an institution or by reviewing the outcome or both. In the light of the many standards, guidelines and procedure, including practices of the universities in the region, the present study suggest certain criteria. It is well established that the quality assurance be based on set policy, procedures and its strategy for rigorous implementation, more important is that quality assurance or criteria setting agency should have formal status and statutory powers. At present, due to increasing numbers of universities, competition among institution/universities automatically forces to ensure quality of education, otherwise less or no admission. The quality assurance criteria are derived by different type of accrediting agencies like government apex agency or association of the respective fields or associations provide input to the government accrediting agency to derive standards, guidelines and strategies.

1. *Policy and procedures for quality assurance*
   The policy and procedure for quality assurance should be based on changing learning environment and implementation relating to different type of learning system like service learning, and to ensure the participation of the students and the community, to relate the curriculum to the new generation and community needs. The policy documents should be publicly available and/or it should widely circulate to create awareness to all the stakeholders and implementers.

2. *Periodic review and updating*

   There should be a formal mechanism for periodic review and updating the guidelines on par with the needs of program, changing learning environment and in consonance with the market demands.

3. *Informed assessment to both teachers and students*

   The awareness should be created to students and teachers about the criteria, regulation and procedures, so that, they are complaint with standards, guidelines and strategy.

4. *Ensuring the teaching competence and reflective learning*

   The quality of teaching and learning also depends on the method of teaching-learning adapted in the institution/university, for example, Service learning or engaged learning. The guidelines should be evolved not just based student-teacher ratio and facilities, it needs to provide or derive based on best practices, for the style of teaching –learning adapted by the institution, be reviewed periodically on par with the fast changing learning environment, and ensure the corresponding qualification and competence of teachers. It is usually has to be reviewed by external agency based on the reports provided by the institutions.

5. *Learning support systems*

   Institutions should ensure the adequacy of learning resources and facilities, based on the type of learning system adapted in the institution, for each programs. It also is required to ensure that institution collect, analyze and use relevant data properly and accurately in addition to provision of learning materials/ resources and students performance measurement.

6. *Use of standards and procedure between internal and external agency*

   External agency should take into account the effectiveness of internal quality assurance, as it depends on the type of students, local culture and learning styles adapted by the institution. Further, output should be authenticated by the student's experiences.

7. *Analysis of the guidelines, whether it fits the learning style adapted in the institution*

   The analysis should be based on the published criteria and its relevance to the style of learning adapted (example service learning or engaged learning), hence ensure the fitness of the standards and guidelines to the aim and objectives of the specific institution. The reports also should be generated not just to submit to accrediting agency; it also should provide a feedback to the institution for furthering their quality and transparency.

   Majority of the guidelines are addressing the manpower, training, performance appraisal of the students with some important quality indicators. Also they address governance, transparency, resources and management strategy. The coordination and fitness of internal and external quality assurance also need to address the value system in education, civic responsibilities and mechanism for continuity and consonantal development with the fast changing learning environment and learner's attitudes.

9.5   Quality Assurance and Existing Situation in Thailand

Despite facing considerable barriers as diversity of cultures, languages and political systems, a successful collaborative scheme for accreditation of LIS education programs was developed by LIS specialists in Southeast Asia. First suggested in Kuala Lumpur at the 2001 International Conference for Library and Information Science Education in Asia Pacific Region, a special committee of the Congress of Southeast Asian Librarians (CONSAL) was set up to take this forward, with representation from LIS schools in the region. The initiative is intended to establish tiered standards for the recognition, endorsement, and accreditation of LIS schools in order to improve their credibility and comparability. Also in Southeast Asia, a collaborative project between Nanyang Technological University in Singapore and the University of Malaya has resulted in the establishment of a Web portal for LIS education in Asia, including a repository of teaching materials for use by LIS schools throughout the LIS schools in Malaysia, Singapore and Thailand. For example, they have been operating exchange schemes and knowledge-sharing mechanisms regarding the use of new media and digital technologies, which have helped them diversify the range of courses and services that they offer.

9.6   Measures to Be Taken

Accreditation is a mechanism for quality control. Monitoring activities of an organization is necessary if it is to improve. Accreditation has several benefits. A qualified department is more likely to absorb (talented) students. It also helps applicants choose the department at which they want to study. The organizations that are normally in charge of accreditation are scientific or professional societies and associations. National LIS associations have not yet been established or empowered in Asia, and therefore, accreditation of departments rarely happens in Asian developing countries. The lack of quality control results primarily in an imbalance and dissatisfaction among graduates and in the job market. Consequently, graduates lose several job positions because they simply are not qualified for those jobs. From the accreditation viewpoint, there are many causes for the association's weaknesses ranging from ethnical trends and issues to powerful personal relationships, lack of legal regulations, lack of awareness about the role of the associations and the importance of their function, etc.

There should be quality control and monitoring system in place in order to guarantee some basic standards for all LIS schools. No new LIS school should be established without the approval of an accreditation agency. This would partly solve the unemployment problem and will improve the social status and self-esteem of the graduates.

Library and information science education is at the departmental level within faculties (schools). The curriculum structure depends on policies and regulations of higher education institutions affiliated. Only the parent institutions of those schools, in cooperation with the Commission on the Higher Education in the Ministry of Education are responsible for the approval of LIS curriculum. The role of professional organization—The Thai Library Association (TLA)—in influencing the formal education system is quite different from counterparts in many other countries. TLA does not perform any accreditation of LIS schools. The TLA assumed some responsibilities, along with LIS schools, both are responsible for professional development and continuing education of information professionals. However, TLA has not accredited the LIS program.

9.7   Conclusions

LIS has low recognition and has not been regarded on par with other well-known professions. As a result, not many talented students choose LIS as their field of study. To solve these problems setting up limited number of independent LIS schools, establishing or empowering accreditation agencies, flexibility in educational systems, more emphasis on research, developing in-service training, relocating the departments in new faculties, equipping the departments with new facilities, employing new and skillful staff, encouraging collaboration among faculty members and departments, diversifying courses and degrees, updating syllabi in an ongoing manner, taking advantage of IT, and creating and publishing LIS literature in native languages should be encouraged.

In the last 60 years of Library and Information Science education has seen many transitions, contrasts, and contradictions in Thailand. The transition is one of the most welcome and significant developments. As of today, Library and Information Science education is on the threshold of facing new challenges of the new century. Great expectations however are in store to establish its durability and survival in the next millennium. If the departments of Library and Information Science in Thailand are to sustain the challenges, there is a dire need to set global standards in Library and Information Science education at least for the Asian region. The task is stupendous and involves drastic and progressive changes in the curriculum and building the LIS courses in light of the happenings in the International arena, the adoption of modular approach is a way of meeting the present and future needs of a dynamic curriculum. Thus, the education and training programs in Library and Information Science must make a provision to prepare the professionals to assume the proactive role in coping with new technology and the information explosion. In brief, the designed course contents should concentrate in developing knowledge, skills, and tools corresponding to the four basic identified areas of creation, collection, communication, and consolidation. This would facilitate the LIS professionals to execute the greater professional responsibilities in the preset Information Society and Knowledge Society. It is hoped that this approach would serve as a guideline to the future curriculum-designing activities in the developing countries (Butdisuwan & Ramesh Babu, 2013). The library schools should assume the role of leadership and responsibility to produce competent manpower for the present as well as future needs of different kinds of information centers including university libraries. To conclude in the words of Lancaster: "We must shift the focus of our professional concern away from the Library as an institution and towards the skilled professionals, who will become a professional practitioner on par with medical and legal practitioners" (Lancaster, 1983).

In order to realize the vision of the above statement it is imperative that quality assurance is the only step and goal. The globalization and harmonization of education has been influencing the higher educational institutions and the respective governing bodies to strive for the maintenance of quality in the teaching, learning and research in higher education. The quality assurance is an ongoing process, ever changing with new directions and never ending phenomena. However, it is not an end by itself but it is a means to an end.

References

Butdisuwan, S. (2000). *Library and information science education in Thailand: General scenario.* Paper presented at the WISE workshop, organized by Nanyang University of Technology, Singapore.

Butdisuwan, S., & Gorman, G. E. (2002). Library and information science education in Thai public universities. *Education for Information, 20*(3/4), 169–181.

Butdisuwan, S., & Ramesh Babu., B. (2013). *LIS Education in Thailand and Tamilnadu: A study in comparison.* In Libraries in the changing dimensions of digital technology. Prof. D Chandran's Festschrift. Delhi: B R Publishing Corporation.

CHEA. (2003). *Statement of mutual responsibilities for student learning outcomes: Accreditation, institutions and programs.* US: Institute for Research and Study Accreditation and Quality Assurance. Council of Higher Education. Washington, DC, http://www.chea.org/pdf/StmntStudentLearningOutcomes9-03.pdf

Enser, P., & Wood, K. (1999). New approaches to the professional accreditation of library and information science education. In: M. Klasson et al. (Eds.) New fields for research in the 21st century: *Proceedings of the 3rd British-Nordic Conference on Library and Information Studies* (pp. 19–26). 12–14 April 1999, Boras, Sweden.

Khoo, C., Majid, S., & Chaudry, A. S. (2003). Developing an accreditation system for LIS professional education programmes in Southeast Asia: Issues and perspectives. *Malaysian Journal of Library and Information Science, 8*(2), 131–149.

Lancaster, F. W. (1983). Future librarianship: Preparing for an unconventional career. *Wilson Library Bulletin, 57*(9), 747–753.

Office of the National Education Standards and Quality Assessment. (2001). *Guideline for external assessment for higher education office of the national education standards and quality assessment (2001). Guideline for external assessment for higher education.*

Premsmit, P. (1999). *Library and information science education in Thailand.* In Libraries and librarianship in Thailand: From stone inscription to microchips, IFLA '99 National Organizing Committee, Bangkok, 71–76.

Ruksasuk, N. (1999). *Library and information science distance education in Thailand in the next decade.* In 65th IFLA Council and General Conference, 1999. Retrieved September 30, 2012, from www.ifla.org/IV/ifla65/papers/090-104e.html

Saccanand, C. (1999). Distance education in library and information science in Asia and the Pacific region. *IFLA Journal, 25*(2), 97–100.

Chapter 10
LIS Education and Quality Assurance System in Asia Pacific: Indonesia

L. Sulistyo-Basuki

10.1 Background

Quality assurance is defined as a planned and systematic review of process of an institution or program to determine that acceptable standards of education, scholarship, and infrastructure are being maintained and enhanced (CHEA, 2003). A literature survey toward quality assurance in library and information science (hereafter called LIS) in Asia yielded few results such as writings of Ameen (2007), Saladyanant (2006), Miwa (2006), Sajjad ur Rehman (2008); papers of Maesaroh and Genoni (2009) and Sulistyo-Basuki (2006) provided little information on that matter. This paper studies the current quality assurance in Indonesia based on the available writings in Bahasa Indonesia (the lingua franca or national language of Indonesia), the available grey literature and interviews.

10.2 Historical Background

The Library and Information Science (hereafter called LIS) education in Indonesia began in 1952 when The Ministry of Education, Teaching, and Culture started a Training Course for Library Employees (Vreede de Stuers, 1953); Rungkat (1997) called it Djakarta Library School. Formerly the study duration was 1 year, later extended to 2 years after the name had been changed to *Kursus Pendidikan Pegawai Perpustakaan* or Librarians' Training Course, then by 1959, to 3 years in order to attract appropriate public service promotion (Dunningham, 1964), the name changed to *Sekolah Perpustakaan* (Library School) (Soemarsidik, 1961). In 1962, it was

L. Sulistyo-Basuki (✉)
University of Indonesia, Depok, Indonesia
e-mail: sbaski@indosat.net.id

M. Miwa and S. Miyahara (eds.), *Quality Assurance in LIS Education: An International and Comparative Study*, DOI 10.1007/978-1-4614-6495-2_10, © Springer Science+Business Media New York 2015

attached to *Fakultas Keguruan dan Ilmu Pendidikan* (Teachers' Training College) University of Indonesia and produced two classes of graduates, i.e., class graduates in 1962 and 1963. In 1964, the department was transferred to *Fakultas Sastra* (Faculty of Letters) Universitas Indonesia. The enrollment was from high school; however, in 1969 the enrollment system was changed, the students must be a holder of *Sardjana Muda* degree (a degree conferred after studying 3 years at tertiary level). By 1970s, there are only two library schools in the nation, University of Indonesia in Jakarta and Bandung Teachers' College in Bandung (West Java). In 1989, the government issued an Act called National Educational Act no 2, 1989 which divided the tertiary education into two mainstreams, i.e., Diploma Program and Academic Program, and later on it was replaced with the new 2003 National Educational System Act.

10.3   The Present Educational Systems

Based on the 2003 National Educational System Act, the tertiary education consists of academic, vocational and professional education. The academic program consists of Undergraduate program (*Program Sarjana*), Graduate program (*Program Pascasarjana*) and Doctorate program (*Program Doktor.*) (Fig. 10.1), while the

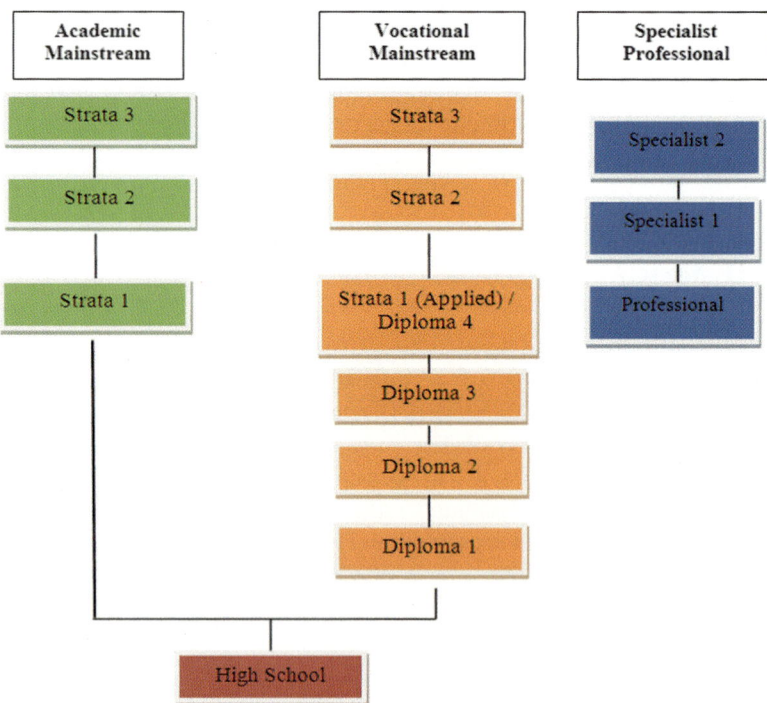

Fig. 10.1 National educational system

Diploma program consists of Diploma 1 through 4, *Magister Terapan* (Applied Master) and *Doktor Terapan* (Applied Doctorate). The Arabic number following the Diploma denotes the students' duration of study at the tertiary education, for example Diploma 3 means that he or she has studied 3 years after high school. After finishing Diploma 4, the student can continue further studies to Strata 2 (in the former 1989 National Education Act it was not allowed) then there is a new program called *Doktor Terapan* which is equivalent to Doctorate degree; the former (*Doktor Terapan*) more or less something like doctorate in profession (for example Education Doctor or Ed. D) while the latter more toward Doctor of Philosophy or Ph.D.

The professional program object is to produce professional in various fields such as nursing, accounting, medical science, etc. The professional mainstream refers to professional world such as physicians, accountants, dentists etc. For academic mainstream, a student finished her or his *Sarjana* which is equivalent with Bachelor degree in many countries. If he or she continued to Strata 2, then she or he or she get a Magister degree, equivalent to Master degree in most of the universities. Continuing to the third program then finishing it she or he got a *Doktor* (doctorate) degree.

The goal of diploma program in higher education institutions is to prepare students to have the ability to apply the science and skills according to each of their department program. Hence the Diploma 1 through 4 have the ability to apply library and information courses, ready for job market, and thus the composition consists of 60 % practice and 40 % theoretical aspects. In contrast, the *Sarjana* program consists of 60 % theoretical aspects and 40 % application ones.

10.4   The Present LIS Education

Based on interviews, long distance calls, the available documents and proceedings and site visits, a directory of library school in Indonesia (Sulistyo-Basuki, 2013a, 2013b) yielded various data. List of operating institutions in various programs are presented in Tables 10.1, 10.2, and 10.3 respectively.

Table 10.1 Diploma programs (vocational mainstream)

No	Institution	City, Province	D2	D3	D4
1	Universitas Terbuka (Open University)	Jakarta	V	V	V[a]
2	Universitas Indonesia	Jakarta		V	
3	Universitas YARSI	Jakarta		V	
4	Universitas Islam Negeri (State Islamic University) Syarif Hidayatullah	Jakarta		V	
5	Universitas Padjadjaran	Bandung (West Java)		V	
6	Universitas Islam Nusantara	Bandung (West Java)		V	
7	Universitas Jenderal Soedirman	Purwokerto (Central Java)		V	
7	Universitas Diponegoro	Semarang (Central Java)		V	
8	Universitas Gadjah Mada	Yogyakarta (Yogyakarta)		V	

(continued)

Table 10.1 (continued)

No	Institution	City, Province	D2	D3	D4
9	Universitas Islam Negeri Sunan Kalijaga	Yogyakarta (Yogyakarta)		V	
10	Universitas Sebelas Maret	Solo (Central Java)		V	
11	Universitas Airlangga	Surabaya (East Java)		V	
12	Universitas Negeri Malang	Malang (East Java)		V	
13	Universitas Udayana	Denpasar (Bali)		V	
14	Institut Agama Islam Negeri Antasari	Banjarmasin (South Kalimantan)		V	
15	Universitas Sam Ratulangie	Manado (North Sulawesi)		V	
16	Universitas Islam Negeri Alauddin	Makassar (South Sulawesi)		V	
17	Sekolah Tinggi Ilmu Sosial dan Ilmu Politik Petta Baringeng Soppeng	Soppeng (South Sulawesi)	V		
18	Universitas Muhamadiyah Mataram	Mataram (East Nusa Tenggara)		V	
19	Institut Agama Islam Negeri Ar-Raniry	Banda Aceh (Aceh)		V	
20	Universitas Sumatera Utara	Medan (North Sumatera)		V	
21	Institut Agama Islam Negeri Imam Bonjol	Padang (West Sumatera)		V	
22	Universitas Lancang Kuning	Pekanbaru (Riau)		V	
23	Universitas Bengkulu	Bengkulu (Bengkulu)		V	
24	Universitas Negeri Padang	Padang (West Sumatera)		V	
25	Universitas Haluoleo	Kendari (Southeast Sulawesi		V	
26	Institut Agama Islam Negeri Antasari	Banjarmasin (South Kalimantan)		V	
27	Universitas Udayana	Denpasar (Bali)		V	
28	Universitas Negeri Malang	Malang (East Java)		V	

Source: Sulistyo-Basuki (2013a, 2013b) *Direktori sekolah perpustakaan* (Directory of Indonesian school libraries, unpublished paper); Asosiasi Penyelenggara Pendidikan Tinggi Ilmu Perpustakaan. Indonesia *Daftar anggota* (2013)
[a]Archival Studies

Table 10.2 Undergraduate programs (Strata 1)

No	Institution	City (Province)
1	Universitas Indonesia	Jakarta (Jakarta)
2	Universitas YARSI	Jakarta (Jakarta)
3	Universitas Islam Negeri Syarif Hidayatullah (State Islamic University)	Jakarta (Jakarta)
4	Universitas Padjadjaran	Bandung (West Java)
5	Universitas Islam Nusantara	Bandung (West Java)
6	Universitas Diponegoro	Semarang (Central Java)
7	Universitas Kristen Satya Wacana	Salatiga (Central Java)
8	Universitas Islam Negeri Sunan Kalijaga	Yogyakarta (Yogyakarta)
9	Universitas Airlangga	Surabaya (East Java)
10	Universitas Brawijaya	Malang (East Java)
11	Universitas Islam Negeri Alauddin	Makasar (South Sulawesi)
12	Universitas Sam Ratulangie	Manado (North Sulawesi)
13	Institut Agama Islam Negeri Ar-Raniry	Banda Aceh (Aceh)
14	Universitas Sumatera Utara	Medan (North Sumatera)
16	Universitas Lancang Kuning	Pakanbaru (Riau)
17	Institut Islam Negeri Raden Patah	Palembang (South Sumatera)

(continued)

Table 10.2   (continued)

No	Institution	City (Province)
18	Universits Pendidikan Indonesia	Bandung (West Java)
19	Universitas Wijaya Kusuma (Surabaya)	Surabaya (East Java)
20	Universitas Terbuka (Open University)	Jakarta (Jakarta)
21	Institut Agama Islam Negeri Sultan Thaha	Jambi (Jambi)
22	Universitas Sari Mutiara	Medan (North Sumatera)
23	Universitas Cenderawasih	Jayapura (Papua)
(24)	(Universitas Hasanuddin)	(Makasar, South Sulawesi)[a]

[a]Archival Management Program
Source: Sulistyo-Basuki (2013a, 2013b) *Direktori sekolah perpustakaan* (Directory of Indonesian school libraries) (Unpublished paper) and also Asosiasi Penyelenggara Pendidikan Tinggi Ilmu Perpustakaan. Indonesia (2013)

Table 10.3   LIS graduate programs

No	Institution	City (Province)	Notes
1	University of Indonesia	Jakarta (Jakarta)	Established 1990
2	Bogor Agricultural University	Bogor (West Java)	Established 2006
3	Padjadjaran University	Bandung (West Java)	Established in 2003
4	Sunan Kalijaga State Islamic University	Yogyakarta (Yogyakarta)	Established in 2008
5	Gadjah Mada University	Yogyakarta (Yogyakarta)	Established in 1996, closed in 2000 then opened again in 2006
6	Alauddin State Islamic University	Makassar (South Sulawesi)	Establish in 2013

Source: Sulistyo-Basuki (2013a, 2013b) *Direktori sekolah perpustakaan*. (Directory of Indonesian school libraries) (Unpublished paper); also Asosiasi Penyelenggara Pendidikan Tinggi Ilmu Perpustakaan. Indonesia(2013)

In Indonesia by the time of writing this paper, there is no doctorate program in LIS (Sulistyo-Basuki, 2013a, 2013b) because of lack of qualified staffs (the government requires two professors and 4 Ph.D.s in LIS while at the present time there is only one professor and it was known only 6 Ph.D.'s in LIS) and other administrative constraints.

10.5   Recognition and Quality Assurance

The quality assurance conducted in various methods such as through Ministry of Education, National Accreditation Agency, Indonesian Library Associations, National Library of Indonesia as described below.

Table 10.4 Courses for
learning to do: expertise to do
(20 credit hours)

No	Course title	Credits
1	Introduction to Library and Information Science	4
2	Telematique	4
3	Research Methods in LIS	4
4	English language	6
5	Indonesian Librarianship	2

Sources: Konsorsium Sastra dan Filsafat (2000)

Table 10.5 Courses for
learning to do: behavior
of doing

No	Course title	Credits
1	Information organization	4
2	Information source and services	4
3	Conservation and preservation	2
4	Library and Information Institution Management	4
5	Information Marketing	2
6	User Studies	2

10.5.1 Ministry of Education and Culture Through Compulsory Curriculum

This system was conducted since 1989 with the edict of Ministry of education and Culture based on 1986 National Educational Act began 1986. In this system, the Ministry issued a nation-wide compulsory course with a total of 144 credit hours. For an undergraduate program, the LIS education required the minimum of 144 credit hours, can be done in 4 years. The maximum credit hours are 160 h. In this system, the Library schools almost have options but to follow the curriculum as the schools have an option to develop optional courses for only 16 credit hours (the difference between the required 144 credit hours to maximum 160-credit hours. Each course is limited to only 2 credit hours. Between 1986 and 1998, there were 2 Ministry issues on compulsory courses. The year 1998 marked the end of President Suharto administration whose policy had strong centralization administration including the curriculum (Sulistyo-Basuki, 2006).

After 1998 (called era of Reformation, after the downfall of President Suharto) there were strong demand from the education field to revise the centralized curriculum. In LIS education, this was followed by a national curriculum designed by the Directorate General of Higher Education (DGHE). It was issued in 2000s which established Consortium for Culture & Philosophy whose domain also covers LIS. The Consortium used the four pillars of education as proposed by the UNESCO-established International Commission on Education chaired by Jacques Dellor (1996). The pillars of learning to do consist of expertise to do and behavior of doing. The course of expertise to do consist of 20 credits as presented in Table 10.4.

Group of courses on behavior to do something consist of 18 credit hours as shown in Tables 10.5 and 10.6.

Learning to live together, learning to live with others need 14 credit hours.

Table 10.6 Courses for learning to live together

No	Course title	Credit
1	Communication	4
2	Library cooperation and information networks	2
3	Professional Ethics	2
4	Psychology of Information User	4
5	Library practice	2

For courses belonging to principle of learning to be, DGHE stated that there are 6 credit hours, compulsory for new students and applies nation-wide for all tertiary education. The course consist of *Pancasila* (the state ideology) 2 credit hours, Bahasa Indonesia or Indonesian language 2 credit hours and *Ketahanan* Nasional (National Resilience) 2 credit hours. Hence total required courses for LIS education at undergraduate level is no more than 52 credit hours giving enough space for LIS administrator to developed their own curriculum. It is a far cry from the New Order era, when the Ministry of Education issued a nation-wide LIS compulsory curriculum with 132 out of 144 credit hours, leaving not enough space for the LIS education administrator to maneuver enough elected courses.

In 2010, the national curriculum names changed to "Core curriculum" and still compulsory for all undergraduate LIS education. It was accepted as the core curriculum for undergraduate program as an answer to critics who doubted the basic required competencies as the undergraduate programs come under various faculties (Table 10.7) and as a guide for the library school operators to develop their own courses.

10.5.2   Accreditation Through National Accreditation Agency

The Indonesia's 1945 Constitution required the Government to have and undertake a national instructional system regulated in a national act. The National Educational Act No. 2, 1989 on National Education System, which is valid since March 27, 1989, and renewed with the National Educational Act No. 20, 2003 on National Education System, is the fulfilment of the constitutional requirement. The Act regulates the whole system of national education, including higher education institutions and study program accreditation system. The accreditation system is regulated in Chapter XVI of the National Act on Evaluation, Accreditation, and Certification, (Article 60 and Article 61). The government established an accreditation agency called *Badan Akreditasi Nasional Perguruan Tinggi* (National Accreditation Agency-Higher Education); it is a non-structural agency under the Minister of National Education, which is initially regulated by the Education and Culture Ministerial Decree No. 0326/U/1994 on The National Accreditation Agency for Higher Education (*Badan Akreditasi Nasional Perguruan Tinggi* commonly shortened as BAN-PT), which was renewed by the Education and Culture Ministerial Decree No. 187/U/1998 on The National Accreditation Agency for Higher

Table 10.7 The position of undergraduate studies at various institutions

Faculty or school supervising the LIS undergraduate programs	Name of institutions
Faculty of Letters	Universitas Diponegoro, Universitas Sebelas Maret
Faculty of Humanities	Universitas Indonesia; Universitas Sumatera Utara, Universitas Lancang Kuning, Universitas Diponegoro
Faculty of Computer Science	Universitas YARSI
Faculty of Information Technology	Universitas Kristen Satya Wacana, Universitas YARSI
Faculty of Education	Universitas Pendidikan Indonesia
Faculty Adab (Culture)	Universitas Islam Negeri Sunan Kalijaga, Institut Agama Islam Negeri Ar-Raniri, Universitas Islam Negeri Alauddin
Faculty of Communication	Universitas Padjadjaran, Universitas Islam Nusantara
Faculty of Administration	Universitas Brawijaya
Faculty of Social and Political Sciences	Universitas Terbuka (Open University), Universitas Airlangga, Universitas Sam Ratulangi, Universitas Wijaya Kusuma (Surabaya)
	Note: Archival Management program at Universitas Hasanuddin also at the Faculty of Social and Political Sciences
Faculty Adab (Culture) and Humanities	Universitas Islam Negeri Sunan Kalijaga, Universitas Islam Negeri Syarif Hidayatullah;, Universitas Islam Negeri Sultan Alauddin, Institut Agama Islam Negeri Raden Patah
Faculty Adab, Sastra dan Kebudayaan Islam (Faculty of Civilization, Islamic Literature and Culture) [sic]	Institut Agama Islam Negeri Sultan Thaha
Unknown	Universitas Cenderawasih

Source: Sulistyo-Basuki (2013a, 2013b); Asosiasi Penyelenggara Pendidikan Tinggi Ilmu Perpustakaan. Indonesia (2013)

Education (*Badan Akreditasi Nasional Perguruan Tinggi*—BAN-PT), then by the Education and Culture Ministerial Decree No. 118/U/2003 on The National Accreditation Agency for Higher Education (*Badan Akreditasi Nasional Perguruan Tinggi*—BAN-PT), and finally replaced by the National Education Ministerial Regulation no 28, 2005 on The National Accreditation Agency for Higher Education (*Badan Akreditasi Nasional Perguruan Tinggi*—BAN-PT) based on the National Education Ministerial Regulation no 6, 2000.

The National Accreditation Agency for Higher Education has just newly introduced a new Quality Assurance system, a modification of the old system with 14 criteria into only 7 criteria/standards. The old 14 criteria/standards have been consolidated nowadays into a more systematic model adapting and combining the QA model of European Foundation of Quality Management (EFQM) and Malcolm Baldrige's model.

The seven new criteria/standards are (1) Vision, Mission, Objectives, and Strategy, (2) Governance, Leadership, Management, and Quality Assurance System, (3) Students (including students' affairs) and Graduates, (4) Human Resources Management (Faculty and staff members), (5) Curriculum, Learning Approach and

Table 10.8 LIS accreditation for Diploma 2 and 3

No	Institution	Province	Program	Results
1	Institut Agama Islam Negeri (IAIN) Ar-Raniry in Banda Aceh	Aceh	Diploma 3	C
2	Universitas Sumatera Utara in Medan	North Sumatera	Diploma 3	B
3	Universitas Bengkulu	Bengkulu (Bencoolen)	Diploma 3	B
4	Universitas Lampung in Bandar Lampung	Lampung	Diploma 3	B
5	Universitas Islam Negeri (UIN) Sunan Kalijaga in Yogyakarta	Yogya-karta	Diploma 3	B
6	Universitas Diponegoro, in Semarang	Central Java	Diploma 3	B
7	Universitas Sebelas Maret in Surakarta	Central Java	Diploma 3	B
8	Universitas Airlangga in Surabaya	East Java	Diploma 3	B
9	Universitas Muhammadiyah Mataram in Mataram	West Nusa Tenggara	Diploma 3	C
10	Sekolah Tinggi Ilmu Sosialdan Ilmu Politik Petta Baringeng in Soppeng	South Sulawesi	Diploma 2	C
11	Institut Agama Islam Negeri Imam Bonjol in Padang	West Sumatera	Diploma 3	C
12	Universitas Negeri Padang in Padang	West Sumatera	Diploma 3	C
13	Institut Agama Islam Negeri Antasari in Banjarmasin	South Kalimantan	Diploma 3	C

Source: (2013) (National Accreditation Agency-Higher Education (Badan Akreditasi Nasional Perguruan Tinggi)

processes, and Academic Atmosphere, (6) Finance, Facilities, and Infrastructures, and Information/ICT Management, (7) Research, Community services, and Collaborations. This new set of accreditation standards framework has been introduced to the stakeholders through a massive dissemination program in 14 provincial capital cities of Indonesia from July to November 2008 and has been implemented for undergraduate study program accreditation since April 2009. A similar approach has been sent to develop and improve the instruments for Diploma, Graduate, and Professional Study programs accreditation. These new accreditation instruments including the academic papers, manuals, guidelines, etc. have just been finished in August 2009 and are under a massive dissemination in the month of October 2009 A field visit to National Accreditation Agency for Higher Education yields some results for LIS education (Table 10.8)

The accreditation is for 5 years, afterward should be renewed, When the data was processed, Universitas Sumatera Utara, Institut Islam Negeri Bonjol, Universitas Islam Negeri Sunan Kalijaga in Yogyakarta, Universitas Diponegoro, Universitas Airlangga, Sekolah Tinggi Ilmu Sosialdan Ilmu Politik Petta Baringeng have submitted complete materials for accreditation for the year 2014. Apparently only 50 % of the existing Diploma programs have submitted their materials for accreditation, while the remaining institutions are preparing the accreditation materials for 2014. The accreditation result for undergraduate program is listed at Tables 10.9 and 10.10.

The accreditation results is used for the university to promote its institution, for attracting bright students, inviting research funds and also a prestigious achievement for the universities; hence the A grade is better than B while grade B is higher

Table 10.9 Strata 1 (Sarjana) program accreditation

No	Institution	Province	Result
1	Institut Agama Islam Negeri (IAIN) Ar-Raniry in Banda Aceh	Aceh	C
2	Universitas Sumatera Utara in Medan	North Sumatera	A
3	Universitas Indonesia	Jakarta	A
4	Universitas Islam Negeri Syarif Hidayatullah	Jakarta	B
5	Universitas YARSI	Jakarta	B
6	Universitas Islam Nusantara in Bandung	West Java	C
7	Universitas Padjadjaran in Bandung	West Java	A
8	Universitas Islam Negeri Sunan Kalijaga in Yogyakarta	Yogyakarta	B
9	Universitas Diponegoro in Semarang	Central Java	B
10	Universitas Airlangga in Semarang	East Java	B
11	Universitas Wijaya Kusuma in Surabaya	East Java	B
12	Universitas Islam Negeri Alauddin in Makasar	South Sulawesi	C
13	Institut Agama Islam Negeri Sultan Taha in Jambi	Jambi	B

Source: Badan Akreditasi Nasional Perguruan Tinggi (2013)

Table 10.10 Graduate programs

No	Institution	Province	Result
1	Universitas Indonesia[a]	Jakarta	B
2	Universitas Padjadjaran[b], Bandung	West Java	A
3	Institut Pertanian Bogor (Bogor Agricultural University), in Bogor	West Java	n.a.
4	Universitas Islam Negeri Sunan Kalijaga in Yogyakarta	Yogyakarta	n.a.
5	Universitas Gadjah Mada	Yogyakarta	n.a.
6	Universitas Islam NegeriAlauddin, in Makasar[c]	South Sulawesi	n.a.

Source: Badan Akreditasi Nasional Perguruan Tinggi (2013)
[a]The accreditation expired in 2013 and currently Universitas Indonesia submitted materials for accreditation 2014
[b]Quoted from its Web site, not available from National Accreditation Agency
[c]Established in academic year 2013/2014, has not submitted its accreditation requirements

than C and so on. So it is not a unique occasion when an LIS department uses its accreditation result for its promotion to recruit new students.

Based on the current Ministry of Education and Culture regulation, a university could open its program first, and then 2 years later submit its accreditation request.

Indonesian Library Association
The Indonesian Library Association (ILA) or *Ikatan Pustakawan Indonesia* (IPI) in Bahasa Indonesia, was founded in 1973 as a merging between three former associations such Association of Indonesian Librarian and Documentalists, Association of Special Libraries and Yogyakarta Library Associations. In its constitution, the Indonesian Library Association does not mention accrediting library schools as one of its activities (*Ikatan Pustakawan,* 2013). It is quite different from its counterparts in North America or UK. Many librarians especially those who graduate from library schools in UK, North America or Australia put their hopes to

ILA on accrediting library education. However, such hope is yet implemented as the Indonesian Library Association board members are almost 100 % National Library of Indonesia officials that prompted Pendit called them as "red tape organization" referring to the car used by the government officials must use red plate'[sic].

In 2007, the operator of LIS educations established an association called *Asosiasi Pendidikan Tinggi Ilmu Perpustakaan dan Informasi* (www.perpustakaan.org) or Indonesian Library and Information Science Education Association (formerly The Association of Library Science Higher Education) with Ms Wina Erwina (Universitas Padjadjaran) as its president. In a telephone interview on May 13, 2013, she said that the association has a plan for quality assurance especially for undergraduate programs; however, it has not been implemented yet. The association held its first meeting in Surabaya, East Java in December 2013.

National Library of Indonesia (NLI).
NLI has no power in accrediting the library schools as it is beyond its authority; all LIS programs are operated by the Ministry of Education and Culture and some under the Ministry of Religious Affairs. In fact, MLI, through its Center for Library Education and Training, confuses the library education system by producing librarians from its center. These librarians can be a high-school graduate or an undergraduate program alumnus who has been working in government libraries for a certain time. When the high-school-based employees attend the NLI Center for Library Education and Training course for 628 h or approximately 3 months, they are librarians; although the proper name is *pustakawan fungsional* or functional librarian and eligible to work at the libraries. For a non-LIS-undergraduate degree holder, she or he can attend the library training center for less than 3 months, and at the end he/she claims himself/herself as a librarian! This system has been criticized by LIS educational institutions as humiliating the profession, producing "instant librarians" who know nothing! The impact is felt at the job market. The "instant librarians" are mostly working at the government public libraries at province, municipal, and regency levels, but not at the school, academic, or special libraries or private corporations. The membership of professional organization (read Indonesian Library Association) mostly are "instant librarians," just a small portion are true professional librarians from formal LIS education institutions. The idea of instant librarians was justified by lack of trained librarians submitted by National Library of Indonesia directorship in mid-1980s; however, the instant system has been in existence since mid-1980s and still operating, regardless of the existence of formal library education. The controversy still continues until present times.

10.6   Certification of Librarians

The emerging issue in Indonesia in mid 2013 is the issue of certifying the librarians. This topic is indirectly related with quality assurance on LIS education. The certification was administered by the Ministry of Manpower and Transmigration (Menteri Tenaga Kerja, 2012), with its sub-organ *Badan Nasional Sertifikasi Profesi*

(Professional Certification Authority Agency), which sets up competency standard, defined as a description of capability of person covering the aspect of knowledge, skill, and attitude (Abdurahman, 2012). The professional certification is issued by Professional Certification Body (PCB) after someone passed the professional exams, covering knowledge, skills, and attitude. This certificate valid only for 3 years, afterward one must pass the exams again. Right now the professional examination is free of charge, but later on one must pay.

10.7   Analysis

On core undergraduate LIS curriculum developed by Consortium Literature and Philosophy (2000), it was agreed through an acclamation in workshop for curriculum, however, in the same workshop during the discussion it was found that the majority of LIS schools copied their curriculum each other and they don't developed curriculum based on their national and local needs. For example a library school in Acheh (North Sumatera) should develop its special need for local needs, such as bibliography on Aceh, manuscripts, local-content-related courses.

Tammaro (2005a, 2005b) wrote about the three models for the recognition of formal qualifications and quality assurances of long learning, vocational education and training in LIS, consisting of program accreditation model, individual lifelong learning, and vocational education and training.

For Indonesia the model is a combination of program accreditation model and individual lifelong learning. The accreditation model is considered appropriate because it requires a formal academic qualification as a basic entry level into the profession and the accreditation is focused on LIS school program. For Indonesia, the entry level depends on the library types, for example entry level for primary school library is Diploma 2 (Fig. 10.1), for junior high school is Diploma 3 while for Senior high school can be Diploma 4 or Bachelor degree (in Indonesia equal to Sarjana or Strata 1). The same requirement also applies to public libraries; however, the heads of public libraries in almost all provincial, municipal, and regency public libraries are dominated by the non-career librarians because they are mostly political appointees (Sulistyo & Agus, 2011, 2013; Sulistyo-Basuki & Irhamni, 2012) For the academic libraries, the entrance level is usually Strata 1 and 2 while for the director of an academic libraries the required level is Strata 2 or master's degree (SNI 7330:2009). The Center for Education and Training, National Library of Indonesia, is producing librarians, mostly civil servant, by training the would-be-librarians in short time, i.e., 628 h face-to-face meetings or equal of 3 months training. This short path to be a librarian has been criticized by formal LIS education programs as producing instant librarians which learns everything but knows nothing! In Jakarta, in the premise of National Library of Indonesia buildings, the 3-month library course was conducted from Monday to Friday, from 8 a.m. to 8 p.m.

[sic]. The other impact toward Indonesian librarianship is the fact that the alumnae of the Center for Education and Training of NLI mostly joined the Indonesian Library Association that prompted Pendit (2003) to call the association as red plate librarians, an allusion to compulsory red-plate usage for government owned cars! (Pendit, 2003).

The individual lifelong learning through various activities is conducted by private organizations, Indonesian Library Association, National Library of Indonesia and library schools; however, the points for continuing professional development are very low. For example if a librarian attends an international seminar held in Indonesia, the point for the promotion is only 1 point, regardless of how scholarly the meeting is.

The private organization and foundations which conducted lifelong learning usually utilized the corporate social responsibility (CSR) funds as required by the government. CSR was defined as the commitment of business to contribute to sustainable economic development, working with employees, their families, the local community, and society at large to improve their quality of life. An example of such foundation is the Surabaya (East Java)-based *Yayasan Pengembangan Perpustakaan Indonesia* which developed various community libraries in some parts of Indonesia (PustakaIndonesia.org) while banking's CSR empowerment can be found in Irawati & Rachman's paper (Irawati & Racaahman, 2012)

By the time of writing this paper (August 2013), the librarian certification is on the way, as shown by the establishment of Professional Certification Body by the Director of National Library (Perpustakaan Nasional, 2012); albeit any organization could set up a Professional Certification Body as long as it is recognized by the Ministry of Manpower. Right now; however, in various social media (JP-APTIKS@ yahoogroups.com; ics@yahoogroups.com and private Web sites), it was heavily criticized as a humiliation for professional librarianship and met resistance from graduates of LIS formal educational institutions. Some librarians even suggested that the National Library of Indonesia should focus on its primary function in collecting and publication in various formats and media as well as developing *Bibliotheca Indonesiana* (special collections on Indonesia regardless the origin of the library materials), not mingling with librarian certification since the certification should be more focused on skilled and trained personnel but not professional. Other well-developed profession associations such as Indonesian Medical Association, Indonesian Engineers Association, etc. have issued no comments yet on the certification issues.

10.8   Conclusion

The best way to implement quality assurance for LIS education in Indonesia is through accreditation agency as it is the most important objective (at least until present time) in guaranteeing quality assurance.

10.9   Suggestion

It is suggested that the National Library of Indonesia, Center for Education and Training should focus on lifelong learning by providing the latest issue and technology in LIS, abandoning its practice producing instant librarians. It has more destructive impact than its benefit.

Increasing efforts to make Indonesian Library Association as an independent association, detached from the National Library of Indonesia in terms of financial dependability, staffing, and works.

The present existing LIS graduate school should work together to establish a national doctorate program as Indonesia is the only nation in Southeast Asia that has no doctorate program.

The LIS education institution should issue its own certificate on competency automatically through its degree, not through Professional Certification Body as a librarian is a professional not a labor or manual worker.

References

Abdurahman, S. F. (2012). *The new paradigm of Indonesian human resources development strategy*. Country Report for the Roundtable meeting of Quality Assurance Agencies of the Organization of Islamic ConferebceMemberCountries. www.mqa.gov.my/aqaaiw/Country Report/Indonesia/Indonesian National Accreditation Agency for Higher Education-2.pdf (Download on September 5, 2013)

Ameen, K. (2007) *Issues of Quality Assurance (QA) in LIS Higher Education in Pakistan*. South Asian Libraries & Information Networks. http://punjabiuniversity.ac.in/pbiuniweb/pages/dlis/salin/kanmalammen.htm

Asosiasi Penyelenggara Pendidikan Tinggi Ilmu Perpustakaan. Indonesia (ATIPI). (2013). Daftar anggota ATIPI (ATIPI Member List) (Indonesian Library and Information Science Education Association)

Badan Akreditasi Nasional-PerguruanTinggi. (2013). *Akreditasi Ilmu Perpustakaan*. http://123libraries.files.wordpress.com/2012/10/akreditasi ilmuperpustakaan

CHEA. (2003). *Statement of mutual responsibilities for student learning outcomes: Accreditation, institutions and programs*. Washington, D.C: US Institute for Research and Study Accreditation and Quality Assurance, Council of Higher Education.InAnna Maria Tammaro. (2005). *Report on quality assurance models in LIS programs*. The Hague: IFLA. http:www.ifla.org/VII/s23.index.htm

Dellor, J. (1996). *Treasure within: Report to UNESCO of the International Commission on Education for Twenty-first century*. Paris: UNESCO.

Dunningham, A.G.W. (1964). Library development in Indonesia. Paris:UNESCO

IkatanPustakawan Indonesia. (2013). *Anggaran Dasar dan Anggaran Rumah Tangga serta Kode Etik Ikatan Pustakawan Indonesia*. Jakarta: Ikatan Pustakawan Indonesia 2012–2015.

Irawati, I., & Racahman, J. B. (2012). Library empowerment program through corporate social responsibility: a study of banking at University of Indonesia. In L. Sulistyo-Basuki et al. (Eds.), *Proceedings of the 15th CONSAL Meeting and General Conference*. (pp 867–877) Jakarta: National Library of Indonesia

Konsorsium Sastra dan Filsafat. (2000) *Kurikulum nasional*. Jakarta: Directorate General of Higher Education. (Unpublished) (National curriculum)

Maesaroh, I., & Genoni, P. (2009). Education and continuing professional development for Indonesian academic librarians: a survey. *Library Management, 30*(809), 524–538.

MenteriTenaga Kerja danTransmigrasi. (2013). *Keputusan tentang Penetapan rancangan Standar Kompetensi Kerja Nasional Indonesia SektorJasaKemasyarakatan, Hiburan dan Perorangan Lainnya Bidang Perpustakaan menjad Standar Kompetensi Kerja Nasiona lIndonesia.* Jakarta: PerpustakaanNasional. (Ministry of Manpower and Transmigration decree on national work competency, public service sector, entertainment and other individual in libraries to be Indonesian National Work Competency Standard).

Miwa, M. (2006). Trends and Issues in LIS Education in Asia. *Journal of Education for Library and Information Science, 47*(3), 167–180.

Pendit, Putu Laxman. (2003). *Profesionalisme pustakawan pelat merah: analisa kritis tentang hubung antara Ikatan Pustakawan Indonesia dan Perpustakaan Nasional Republik Indonesia.* (Red-plate librarians' professionalism: a critical analysis of relationship between the Indonesia Library Association with National Library of Indonesia)

Perpustakaan Nasional. (2012). *Keputusan Kepala Perpustakaan Nasional nomor 73 tahun 2012 tentang Lembaga Sertifikasi Profesi Pustakawan.*

Rungkat, T. (1997). *Education and training for librarianship in Indonesia, 1945-1984.* Melbourne: Ancora Press.

Sajjadur Rehman. (2008). Quality Assurance and LIS Education in the Gulf Cooperation Council (GCC) Countries. New Library World, *109*, 7/8, 366–382.

Saladyanant, T. (2006). Quality assurance of information science program: Chiang Mai University. Proceedings of the Asia-Pacific Conference on Library & Information Education & Practice 2006 (A-LIEP 2006), Singapore, 3–6 April 2006 (pp.432–435). Singapore: School of communication & Information, Nanyang Technological University.

Sulistyo-Basuki, L. (1994). *Periodisasi perpustakaan Indonesia.* Bandung: Remadja Rosdakarya (Historical Period of Indonesian libraries)

Sulistyo-Basuki, L. (2006). Political Reformation and Its Impact on Library and Information Science Education and Practice: A case Study of Indonesia During and Post-President Soeharto Administration. In C. Khoo, D. Singh & A.S. Chaudry (eds). *Proceedings of the Asia-Pacific Conference on Library & Information Education & Practice 2006 (A-LIEP 2006), Singapore, 3–6 April 2006* (pp. 172-179). Singapore: School of Communication & Information, Nanyang Technological University.

Sulistyo-Basuki, L. (2013). Direktori Sekolah Perpustakaan. (Directory of Indonesian library schools). Unpublished paper

Sulistyo-Basuki, L. (2013). A study of the curriculum of Indonesia's existing five graduate LIS programs Paper for Asia-Pacific Conference on Library &Information Education & Practice 2013 (A-LIEP 2013). KhonKaen University, Khonkaen, Thailand, 10–12 July.

Sulistyo-Basuki, L., & Agus. (2011). *Laporan monitoring evaluasi perpustakaan umum di provinsi Sulawesi Selatan.* Jakarta: Perpustakaan Nasional. (Unpublished paper) (Report on evaluation and monitoring of public libraries in the province of South Sulawesi).

Sulistyo-Basuki, L., & Agus (2013). *Laporan monitoring evaluasi perpustakaan umum di provinsi Sumatera Barat.* Jakarta: Perpustakaan Nasional (Report on evaluation and monitoring of public libraries in the province of West Sumatra). (Unpublished paper)

Sulistyo-Basuki, L., & Irhamni. (2012). Laporan monitoring evaluasi perpustakaan umum di provinsi Sulawesi Tengah Jakarta: Perpustakaan, Nasional. (Unpublished paper.) (Report on evaluation and monitoring of public libraries in the province of Central Sulawesi).

Soemarsidik, S. (1961). Indonesian School of Librarianship has 104 students. *Malayan Library Journal, 1*, 27–8.

Tammaro, A. M. (2005a). Recognition and quality assurance in LIS: New Approaches for Lifelong Learning in Europe. *Performance Measurement and Metrics, 6*(2), 67–79.

Tammaro, A. M. (2005). *Report on quality assurance models in LIS programs.* The Hague:IFLA. http://www.ifla.org/VII/s23/index.htm. Download July 23, 2013.

Vreede de Stuers, C. (1953). The first library school in Indonesia. *UNESCO Bulletin for Libraries, 7*, E90–E91.

Yayasan Pengembangan Perpustakaan Indonesia (2013). Pustakaindonesia.org. (Personal communication with Ms Triniyanti. second December 2013).

Part III
Regional Quality Assurance System of LIS Education

Chapter 11
Current Trends of Quality Assurance Models in LIS Education

Anna Maria Tammaro

11.1 Globalisation and Quality Assurance in LIS Education: An Introduction

Quality assurance (QA) is an all-embracing term, referring to an ongoing and continuous process of evaluating (assessing, monitoring, guaranteeing, maintaining and improving) the quality of the higher education system, institutions or programmes (UNESCO CEPES, 2007). QA focuses both on objectives of accountability and quality improvement. The current trends of quality assurance models in Library and Information Science (LIS) education should be understood in an international perspective, such as international dimension, international curriculum, transnational courses.

International dimension is based on the assumption that internationalisation must be considered essential to the mission of all higher education institutions. The focus is primarily on gaining intercultural experiences (Abdullahi, Kajberg, & Virkus, 2007). In addition, it is assumed that by enhancing the international and intercultural dimension of teaching, research and higher education institutions themselves, the quality level of higher education systems will be boosted (Boaz, 1986).

Curricula internationalisation has the aim of adding an international element to the content and delivery of LIS programmes. This category covers a wide variety of cases. The first efforts involved the international harmonisation of LIS curricula. UNESCO (1984) was the first to seriously consider the education of information professionals, which resulted in the emergence of the basic idea of harmonisation, i.e. the design of a harmonised programme (UNESCO, 1998). The most prominent form of curricula internationalisation is the delivery of a programme in a language different from that of the country in which the programme is offered, such as English language taught courses. Experts classify this type of offering as "internationalisation

A.M. Tammaro (✉)
University of Parma, Parma, Italy
e-mail: annamaria.tammaro@unipr.it

M. Miwa and S. Miyahara (eds.), *Quality Assurance in LIS Education:*
An International and Comparative Study, DOI 10.1007/978-1-4614-6495-2_11,
© Springer Science+Business Media New York 2015

at home or IaH" (Abdullahi & Kajberg, 2004). Together with English taught programmes, several types of country comparative and international studies also fall into the category of curricular internationalisation. A further group of international curricula are those jointly delivered by two or more higher education institutions in at least two countries (Tammaro & Dixon, 2003). They are a joint or double degree, with a common curriculum and recognised period of study abroad.

Transnational courses and enrolment of international students represents the specific aim of some LIS schools (Abdullahi et al., 2007), compensating for budget shortcuts and loss of national students. Collaborative or cross-border provision entails delivery of programmes with the help of a licensed foreign tertiary institution and various forms of distance (usually online) education methods.

In this global scenario, international quality assurance of LIS education is regulated by the World Trade Organization (WTO, 1998) General Agreement on Trade in Services (GATS). The World Trade Organisation General Agreement on Trade in Services (WTO-GATS) has approved a multilateral framework that sets general rules for the conduct of international trade in services, including education services. The provisions sets by GATS are relevant to LIS education in two important aspects:

- standards of quality of LIS education
- recognition of qualifications for professionals

The trade scenario started by the WTO-GATS agreement, presents many risks for higher education institutions. Many fear that an unregulated global higher education market will give way to a devaluation of quality standards. In a more demand-driven educational market, standards tend to adapt to the demands of customers. The internationalisation of higher education could also be dangerous for the consumer, if it lacks transparency. To counteract these risks, many guidelines and codes have been developed by international organisations, such as the work done by UNESCO (2002, 2004) and the OECD (OECD, 1996), and in Europe the Bologna Process (BP) (Campbell & Van der Wende, 2000; Tuning, 2004; Tammaro, 2007).

All these international guidelines and codes of practice for quality assurance and recognition of qualifications aim at three broad objectives:

- improve transparency of programmes and qualifications
- stimulate cooperation and mutual recognition between two or more countries
- foster the international cooperation and professional networks for recognition

Transparency: the first objective addresses the transparency of qualifications/levels and structures of programmes. Transparency has to be achieved through common structure of the courses and common systems of recognition, as for example years of study, and tools such as the European Qualification Frameworks. Kajberg and Lorring (2005) evidence the "europeisation" project developd in Europe by EUCLID (association of LIS professors in Europe).

Cooperation between countries: the second objective deals with relationships between countries, which cooperate to agree on common criteria of recognition and quality. There are some notable examples of this cooperation in LIS education (Wozniczka-Paruzel, 2003; Virkus & Hartley, 2003), including the LIS joint courses (Tammaro & Dixon, 2003).

Agreed quality criteria: the third objective refers to internationalisation and quality assurance experiences, in which it is possible to agree on quality guidelines and—possibly—on quality assurance procedures with a leading international body. To this purpose, related to stimulating an international process of quality assurance, the international professional associations should develop guidelines on recognising standards of professional programmes, respecting national sovereignty and avoiding uniformity.

Concerning this third objective, bottom-up consensus-building and voluntary acceptance of shared principles were the favourite procedures used by IFLA Education and Training Section. This chapter describes the work done by the Education and Training Section of IFLA for developing a quality assurance model with internationally agreed criteria (IFLA SET Guidelines by Fang, Nauta, & Fischer, 1987; Dalton & Levinson, 2000; IFLA. Section Education and Training, 2009; Tammaro, 2005).

11.2 Quality Assurance in LIS Education: Issues and Trends

Three approaches to quality assurance have emerged from the analysis of various LIS guidelines and standards (Knox, 2001; Khoo, Majid, & Sattar Chaudry, 2003). The three approaches correspond to different phases of the educational cycle: (1) programme, (2) educational process and (3) learning outcomes.

Programme: this orientation is driven by Government QA Agencies and stresses accountability and consumer protection. The criteria most commonly used in LIS Guidelines assume that learning takes place if institutions provide certain inputs or resources (e.g. curriculum content, limited class size, full-time faculty, student workload, documented policies, equipped classrooms and libraries). Quantitative indicators such as number of students enrolled and drop out rates are also important. Quality is intended as fitness for purposes and value for money.

Educational process: these quality indicators include the major decision areas for teachers, administrators and university quality audits. The assumption is that if the learning and teaching process is well carried out, the success of the education is assured. The monitoring of the educational process is continuous with a combination of self-evaluation and external evaluation. When specifying quality standards, some define minimum requirements and others look for identifying excellence. Industrial standards are often used, such as TQM or EFQM, which usually stress world-class benchmarks and excellence (Harvey, 1995).

Learning outcomes: focus is on explicit and detailed statements of what students learn, such as the knowledge, understanding, skills and abilities. The adoption of a learning outcomes approach focuses on the student achievements, competences and employability. The assessors involved in a learning outcomes approach are professional associations, higher education institutions with the involvement of students' active participation in the assessment. The quality assurance model in this case stresses a transformative concept of quality of learning, and is based on individual assessment. Other quality assurance procedures include the subject benchmarking. Subject benchmark statements set out expectations about standards of honours degrees in broad subject areas. The benchmarking process in LIS is carried out only in the UK (Huckle, 2002).

These different approaches evidence that quality in LIS is a value judgment, differently interpreted by various stakeholders, such as governments, employers, students, administrators and LIS teachers. Because quality is a very subjective concept, it is quite important to identify the accrediting body in order to understand the procedures and purposes of the evaluation as well as to establish the authority and validity of the evaluation. Just as there are many concepts of LIS, there are also many definitions and concepts of the quality of LIS education. In an international framework, LIS education should focus on competition or cooperation? Are LIS schools looking for a core curriculum or innovating the curriculum? does employability constitute a measure of quality of LIS education?

11.2.1   Cooperation or Competition?

Cooperation or competitiveness in LIS education is a problem related to the changing role of universities. The quality of learning is one of the reflections of this difficult equilibrium, choosing between:

- centres of excellence which have the best infrastructures and teachers at their disposal are pursued
- minimum requisites which are established to encourage everyone to try to reach them

Depending on who is leading the quality assurance of the programme, there can be different purposes underlying the process: for example, the administrators assessing the LIS programme can be oriented towards competition for excellence.

A LIS school is competitive when is a Centre of excellence and has resources, such as a pleasant work environment, a dedicated campus and large teaching team with consequent specialisation, their curricula are updated and relevant and this is demonstrated through validation and accreditation process, best demonstrated by the existence of accreditation by professional bodies and students success in the labour market.

While an initial trend of the internationalisation of QA was towards the harmonisation of the LIS curricula and the transparency of the minimum requisites, or what is called the core programme, an apparently opposing trend in the global environment is towards the stimulation of excellence and innovation in the LIS curricula. Competition includes the analysis of aspects of the curricula that appear to attract better-performing secondary school students. Should LIS schools have different specialisations and attract students owing to their quality?

11.2.2 Core Curriculum and Curriculum Innovation

A particular challenge of LIS education today is to address the education of future professionals in a field featuring major change and rapid evolution. A great deal has still to be done to resolve or at least clarify the confusion regarding the basic concepts of LIS discipline.

This is necessary to obtain the academic dignity of the discipline and its specific identity. This identity must be understood not so much as the "core", which remains always the same, but something which is adaptable to different situations and to different historical periods. We must also to understand, and if necessary, contrast competition with other professions, which the advent of the digital age has brought about along with the far-reaching transformation of work procedures and organisation.

This core however does not exist if not in a situation of change. The identity of the profession is to be sought for in continual change: using the best technologies available to efficiently achieve an active role in society. While many focus on the core of the curriculum, or the conservative elements in a curriculum, which do not change in different spaces and times, one should stress on the need of relevance and updating of curricular considering innovation and adaptation to different circumstances more important than the core. This is also the reason why the first efforts to create a common curriculum were abandoned. Burnett and Bonnici (2006) have provided evidence, through the accreditation history of library schools and computer science departments, on how the two major groups in the iSchool movement, Computer Science and Library Science, met and forged alliance.

Why is innovation important? In a period of change, due to recent technological developments, particularly networking and the creation of more sophisticated digital libraries, there exists the need to be a thorough overhaul of the essence of what we, as librarians, deal with, and what we set out to do. Developing and implementing services based on digital information and being a partner in the development of new teaching and learning environments, means that a whole new set of skills will be available in the information services (Chu, 2010). It is therefore necessary to change curricula and priorities, in response to developing new curricula which should reflect the preparation of experts in the Information Society and to facilitating international information management competencies (Myburg & Tammaro, 2013).

The continuous updating of LIS curricula, adapting them to change, is considered an important quality indicator. The impact of the new technologies offers challenges to the profession. How can we resolve them? By using the strategy of the "protected species"? This is the answer given by American professors, in using specialisation (Van House & Sutton, 1996). A second answer which is being implemented is that of generalisation, convergence, a research approach (Audunson, Nordlie, & Spangen, 2003).

Change is the major driver for motivating people to learn, to go in depth in their understanding; however, it is necessary that the LIS schools support and encourage this individual stimulus, teaching learning skills, research methods and reflective practice.

11.2.3 Employability

Professional organisations, already operating in the international labour market, force labour market to responsiveness of LIS quality indicators. Today the market orientation in LIS education prevails (Kajberg, 2002). Employability is the most discussed objective in quality of teaching: it is the aim of successful students, although related to local constraints, often in contrast with internationalisation. The discussion on LIS university education has concentrated for years on a contraposition between theory and practice, between university education and vocational training, which was at the basis of the gap between the profession and the academy. Does a possible gap between LIS community and Society exist?

Although LIS as an academic and educational undertaking has common historical roots related to the need of producing qualified staff for work in libraries, research and education has developed in different directions. If today you meet a person who considers himself/herself an educated librarian, i.e. educated by a college or a university, you do not know if the person in question holds an academic degree and, if so, at which level (bachelor or master) or if he has a vocational diploma not integrated into the system of academic degrees. And if the person in question has for example an MA/MSc, you do not know if he/she has studied LIS for 1 or 2 years, building the MA/MSc upon a bachelor in another subject, or for 5 years, building the MA/MSc upon a bachelor in LIS (Audunson et al., 2003). The issue of employability is connected to the issue of the recognition of qualifications which is not often understood by teachers as very few teachers are familiar with qualification levels.

An LIS school is competitive when its graduates are able to undertake the many, extremely diverse, roles that are required in information work. The profile is not only that of the librarian. The possible work areas will vary considerably. Another significant measure of the a LIS school competitiveness is, producing graduates who are interested in, and committed to, engaging in theory development in the field, and in further research. That is, students should develop a curiosity, and give a sense of meaning to their field. However the most important issue to address is the following: is LIS an academic or vocational discipline?

In summary, there is a strong scepticism about the possibility of common standards in national and international systems in quality assurance of LIS education. In order to reach an agreement upon quality criteria we need the following items:

- a compatible quality assurance systems, especially regarding the setting of threshold standards based on learning acquired (outcomes) rather than on time spent and curriculum content (resources)
- a coordinated approach to quality standards for international education, which raises the question of the recognition of qualifications given by foreign education providers

11.3 IFLA Education and Training Section

The International Federation of Library Associations (IFLA) Education and Training Section (SET) has been active in supporting internationalisation and quality assurance in LIS education long before than the WTO-GATS, with a focus on two activities: core curriculum and equivalency of qualifications. The aim of these efforts was to facilitate the recognition of professional qualifications across national borders and to enhance the quality of LIS education globally.

The IFLA Section published some tools aimed at achieving greater transparency of professional qualifications and establishing standards for assessing the quality:

Guidelines for equivalence and reciprocity of professional qualifications (by Fang et al., 1987). These would provide opportunities for improving the skills of individual students and increasing the quality of the national LIS schools. It was recognised that LIS education would/should be at the tertiary and/or post-tertiary level in most countries;

International Guidelines for equivalency and reciprocity of qualifications for LIS professionals. Draft guidance document for transparency, equivalency and recognition of qualifications (by IFLA. Section Education and Training, 2009). These would provide a framework of principles, standards and procedures to: assure transparency, stimulate cooperation between LIS schools, foster international collaboration for quality assurance of LIS programmes.

Quality assurance models in LIS programmes (by Tammaro, 2005). This survey explores how quality is assured in LIS programmes to stimulate cooperation and transparency.

Guidelines for LIS Educational Programmes (by Smith, K., Hallam, G., Ghosh, S.B. updated 2012). These guidelines assist in determining the programme standards and a core curriculum and are regularly updated.

World Guide to Library and Information Science Education by A. Schniederjürgen (Ed.); updated in 2007. This publication lists all the institutions offering education in LIS worldwide. The 2007 edition includes more than 900 institutions and 1,500 LIS programmes.

11.3.1 Quality of LIS Education

The quality assurance models in LIS schools (Tammaro, 2005, 2006) were investigated in a research project started at the IFLA Berlin Conference in 2003. The objectives of the research were to identify the quality indicators and how LIS education quality was measured and evaluated. The primary purpose of this survey was to gather data from a sufficient number of LIS schools worldwide related to current quality assurance processes, priorities and concerns. The research questions were:

- How can we improve quality in LIS Schools?
- How can we preserve diversity within an international framework of quality assurance?

Table 11.1 Survey of library schools

Regional area	Questionnaires sent	Questionnaires received	% Respondents	Questionnaires received by countries	% Countries
Africa	15	2	4	2	5
Asia	21	7	14	6	13
Europe	33	27	54	27	60
Latin America	33	9	18	7	15
North America	58	5[a]	10	3	7
Total	160	50	100	45	100

[a]US and Canadian Library Schools are all accredited by the ALA; they received a modified version of the questionnaire, asking them if they had an accreditation system other than the ALA

A Library schools' survey questionnaire was sent by e-mail to 160 LIS schools worldwide and an added investigation was prepared for US and Canadian LIS Schools, as they all have ALA/COA accreditation systems. The replies were aggregated according to the five regional areas and, inside such areas, by countries. The countries represented were 45. The USA has been considered as a single country, each state being regulated by the same quality assurance system. The replies received from regional areas are spread as indicated in Table 11.1.

The replies were analysed considering:

- the assessor or accreditor of the programme
- the way to measure performance
- the performance indicators and outcomes

1. *Assessor or accreditor of the program*

The differences from one regional area to another are relevant. Africa is the area where quality assurance seems less attended to, with 50 % of respondents equipped with a quality assurance system (note that only two countries responded to the survey). In North America and Canada the Professional Association model is to the fore, while in Europe the Government Agency model prevails. Asia also shows an organisation for quality assurance, with a Government Agency the most diffused and internal Quality Audit of secondary importance. Asian Professional associations are embarking on the task of evaluation of Library schools. Latin America Library schools (85.7 %) have a quality assurance system, with Government playing the major role in evaluation, while Professional associations (20 %) are less involved.

Unless leading bodies, such as Government Agencies are also awarding and accrediting bodies, does LIS quality fall outside their responsibility? Whilst the leading body can specify the number and range of units for assessment, and give advice on how evidence may be collected, the actual assessment should be left to Professional association agencies and verified by the awarding and accrediting bodies. Assessment agencies include a wide range of organisations, even employers, for example, can set themselves up as assessment centres, both for their own employees and for those of other firms within their occupational area. Specialist consultancies are already emerging for undertaking these activities. Indeed, particular regulations are prescribed as required competences for assessors and verifiers. Who assesses the assessors? Who certificates the verifiers?

2. *Ways to Review Performance.*

The quality assurance process most diffuse in LIS Schools is in four phases:

- periodical evaluation process
- self-assessment
- peer expert site visit
- follow-up report

The process usually takes place every 2–5 years (66 %), with self-assessment and site visit (55 and 52 % respectively), often combined. Differences could be evidenced for the follow-up report, not often produced (only 41 %) and with limited publicity (only 7 % made the report public).

Most of the respondents said that guidelines are followed. Typically the guidelines are part of an accreditation handbook or policy manual that includes a description of the accrediting process, the eligibility requirements, relevant policies that institutions must address in their self-study reports and other documentation developed to assist institutions that are preparing self-study and conducting evaluation and assessment exercises. The policy generally elucidates standards and relates to their application. Most European LIS schools have to follow the guidelines which are given by the Government Agency that are common to all universities and not subject related.

3. *Performance indicators*

The resources and curriculum content design indicators are ranked higher (respectively 66 and 83 % of countries) which is consistent with the fact that input measures are more diffused than others. Quantitative and demographical data on students are also considered quality indicators by 48 % of countries.

Other indicators (21 %) refer to teaching staff quality, e.g. professional experience, academic background, contribution to the professional development, including research productivity, value based education, cultural meetings etc., international activities, teaching materials, academic and service staff.

A regional area review concerning the importance attached to these indicators evidence some differences. For instance, the curriculum design and content are considered the most important indicators by 100 % of respondents from all countries; Europe and Latin America rank the resources indicator at about 80 %.

4. *Ways to look at outcomes*

The outcomes focus is less used compared to input measures. Students are involved in quality assurance in 69 % of countries. Learning outcomes is used by 52 % of countries, at different levels. Other output measures have been indicated (about 14 %) such as: percentage of students working after graduation, approval of work done by students on behalf of library professionals, measure of relevance to the labour market, research and scholarly publication activity and strategic position of the programme within the university.

A regional review of the importance attached to these factors reveals more similarities than differences. The learning outcomes approach is diffused in Asia, Africa and North America (100 %), while students' evaluation is less popular. In Europe and Latin America students' evaluation is preferred, while learning outcomes assessment is less used. It should be noted that the Bologna process in Europe is aiming at placing primary importance on the latter approach and in the future the situation could change. In Latin America, the outcomes based approach is weak, for what concerns both learning outcomes (40 %) and student satisfaction (40 %). In North America the outcomes based approach is very popular and diffused. Other indicators are related to staff teaching evaluation for promotion, percentage of students working after graduation, relevance to labour market, research done by students.

The analysis of data was performed with a view to developing a typology of approaches and understanding the different rationales for assessing quality. In addition, the perceived advantages and disadvantages as well as the costs and benefits of the various approaches were examined. The main finding of the survey has been a quality model, which is based on a taxonomy covering quality criteria/processes/definitions to describe, specify, and understand critical properties, characteristics, and metrics of quality in LIS. Regarding what quality assurance covers in LIS, it can be said that there are more similarities than differences: quality assurance is more focused on resources and curriculum design (respectively 73 and 86 %) than on learning outcomes (59 %) and student evaluation (66 %). Learning outcomes model and subject benchmarking should be encouraged as they can be adapted to the diversity of LIS educational programmes and of different context.

When accrediting a LIS programme the following categories should be evaluated:

- Stated learning objectives and the evidence of their achievements (learning outcomes)
- Design and content of the curriculum
- Assessment of student learning outcomes through exams and/or employee evaluations
- Resources in terms of funding, staff numbers and IT/Library facilities
- Number of students, dropout rates, recruitment
- Effectiveness of teachers, staff qualifications
- Responsiveness to learner backgrounds and preferences, pedagogy
- Support for learning
- Student evaluation of the learning experience

The Learning Outcomes Model focuses on explicit and detailed statements of what students learn: the knowledge, the understanding, skills and abilities which higher education institutions seek to develop and then test. This means passing from a prescriptive QA system to a more descriptive one. This model provides a way of conceptualising the effects or results of the programme education system (content, learning process and outcomes) and the influences on different outcomes. This approach was represented as a paradigm shift from traditional ways of measuring learning, characterised as input approaches (emphasising teaching hours expressed in ECTS and counting resources) to output-focused methodologies using learning outcomes and competences. The emphasis on outcomes moves the criteria for quality from the input (resources, what staff teach) to the outcome (what students will be able to do). The adoption of a learning outcomes approach focuses on the learner and not on the teacher. It promotes the idea of the teacher as facilitator or manager of the learning process and recognises the fact that much learning takes place outside the classroom, without a teacher present.

The outcomes assessment process is not only important for quality assurance: it also enables the lifelong learner, from student to full professional status, to trace his/her progress through the identification and recognition of knowledge and skills acquisition and further training needs (Brine & Feather, 2003). This approach is of more relevance to the labour market, and is certainly more flexible when taking into account issues of lifelong learning, non-traditional learning, and other forms of non-formal educational experiences. The quality assurance model in this case stresses the transformative concept of quality assessment and prescribes methods to measure it.

The complexities of the education process must be borne in mind in applying this Learning Outcomes Model in different countries. Education has a range of direct and indirect effects on individuals and society at large that could be measured. The influences on learning outcomes are potentially vast and include many factors outside the education system. Specification of the framework, i.e. determination of the evaluation framework domains, involves deciding which types or categories of outcomes to measure, as well as which categories of influences on outcomes the indicators framework should cover. Some of these may also be factors on which the intervention of higher education institutions has neither direct nor reasonable influence. For example:

- There are factors outside the learning and teaching process that have an impact on outcomes, such as students' personal characteristics, work environment constraints, labour market characteristics, etc.
- Student participation and responsibility in the learning process must not be underestimated

11.4 Equivalency and Recognition of Qualifications

Dalton and Levinson (2000) conducted a first study for IFLA SET Section on LIS qualifications worldwide, with the goal of determining acceptable criteria and procedures for establishing equivalency of qualifications. Feasibility of different approaches was sought following three different approaches:

- Database of national accreditation criteria: this proved impractical since it was discovered that most of the world did not have accreditation criteria specifically for LIS education

- International expansion of the existing NARIC: since NARIC (National Academic Recognition Information Centres) is limited to EU countries, expanding the database internationally would be an overwhelming task (ENIC-NARIC net, 2008)
- Database of LIS course content and duration: this could include each LIS education institution in the world. They did note the challenges of keeping such a database current as well as recognising that most countries did not have library associations that oversaw the quality of LIS education programmes, which would likely be a significant barrier to the realisation of this third approach

Weech and Tammaro (2008) 8 years later started a second study for IFLA Section Education and Training on "Feasibility of Guidelines for Equivalency and Recognition of LIS Qualifications". The objective of this study was to test the feasibility of guidelines by surveying participants and focusing their opinions in equivalency of educational programmes and qualifications for employment. The study was coordinated with relevant activities of other IFLA sections and groups, including but not limited to CPDWL (Continuing Professional Development & Workplace Learning Section) and the LIS Education in Developing Countries Special Interest Group.

The situation of equivalency and recognition was described evidencing: (1) Professional Qualifications, (2) Professional Associations role, (3) Accreditation and Recognition procedure, (4) Learning outcomes approach, (5) Stakeholders communication.

1. *Professional qualifications*

The findings reported by the research survey suggested that to enter the profession a Bachelor degree is required, whereas a Master's degree is required in only a few countries. Besides, there are countries which require a generic degree, not a LIS degree. For civil servants, additional requirements are: individual certification, a professional or generic exam. For career advancement, in many cases a professional retraining or the completion of a Master's degree is required.

Based on the review of the literature on the subject, it is clear that what is lacking is a uniform basis for assessing equivalent degrees and qualifications internationally. In a small number of countries maintaining LIS education according to the Anglo-American model, there are organisations and/or national bodies that provide a basis for making some comparisons and assessments. But in most of the rest of the world, there are no organisations or national bodies that take on this responsibility. Thus the question remains, what is the feasibility of developing some form of procedures or guidelines that will be applicable internationally?

2. *Professional associations role*

Most of the participants in the survey on "Feasibility of Guidelines for Equivalency and Recognition of qualifications" would like IFLA to assume an active role in stimulating member associations on this issue in their individual countries. A Quality model should be given by IFLA, in order to achieve transparency and facilitate recognition by respondents in Europe (73 %) and by some (50 %) in the USA and (50 %) respondents in Asia. Respondents from Asia and Europe would like IFLA encourage member associations and institutions to establish systems of accreditation and/or certification in their country or region based on a recognised Quality Assurance model. For some (20 %) of the respondents in Europe and (50 %) in Asia, IFLA can/should assume a more active role, endorsing the task of national recognition of individuals.

3. *Accreditation and recognition procedure*

How can IFLA or an international library organisation achieve the task of an active role in QA? Three models were suggested:

- International resource centre on relevant information about LIS education
- International expert committee for the assessment of LIS education on an advisory base
- Learning outcomes to be met by all LIS professionals who wish to have their training recognised internationally

The replies show that many of the respondents would prefer the third approach: a quality model focused on learning outcomes (53 % in Europe). The alternative approaches are, in order of preference: an international resource centre (50 % in Europe) or the international experts committee (40 % in Europe). In comparison with other regional areas, the third approach is the favourite: a quality model focused on learning outcomes was chosen by 50 % in the USA and 50 % in Asia. An international resource centre is preferred by 50 % of Asia and US respondents and the international experts committee is preferred by 50 % of Asia respondents but not considered desirable in the USA.

4. *Learning outcomes*

A different approach consists in focusing on which learning outcomes students should have to be competent professionals. A number of tools apt to substantiate leaning outcomes already exist, such as in Europe the Diploma supplement, the "EuroCV", and others. However, few of these tools are widely known and there is a gap in the LIS Sector for their application. If these are to be used as reference tools across Europe on a systematic basis, much greater awareness about their existence and their potential use and benefits is needed.

Respondents were asked to give their opinion on two different approaches to learning outcomes:

- a subject benchmarking system established by sharing best experiences of LIS schools creating benchmarks to assess quality through a peer review process
- a second approach linking quality assurance of LIS education to the assessment of LIS programmes by professionals who successfully completed the courses at each school. (Tammaro, 2005, p. 19)

The first approach of subject benchmarking was indicated as being preferred, respectively by 60 % of respondents in Europe, 100 % in Asia and 50 % in the USA. The second approach, associating QA with assessment of students' achievements was indicated as 50 % in Asia and the USA and 53 % of the preferences in Europe.

5. *Stakeholders communication*

As to the recognition of qualifications, stakeholders' collaboration seems to be difficult to obtain in the countries where Library Associations are not involved in quality assurance. Library Associations seem sometimes - according to the perceptions of some respondents—to not understand innovation and their conservative approach lowers the level of the profession to practicalities. On the other hand, employers' and the labour markets' relationship with Library Schools are improving and internship is playing a crucial role in learning in Library Schools. In the UK the use of students' placement has been encouraged in big industries, and not only in libraries. Most of the participants would preferably see IFLA, or other professional associations assume an active role in prompting member associations to be active on this issue in their country.

11.4.1 IFLA SET Quality Model

The "International Guidelines for Equivalency and Reciprocity of Qualifications for LIS Professionals" by Terry Weech and Anna Maria Tammaro were prepared and published at the end of the feasibility study in 2009. IFLA SET recognised that a "Quality model" should be given to achieve transparency and facilitate recognition. This model aims to determine:

- best measures of quality assurance of LIS education programmes, as evidenced in the survey and the feasibility study, in the judgement of LIS professionals and LIS faculty worldwide
- acceptable criteria and procedures for establishing equivalency and reciprocity of LIS Professionals qualifications

In those countries with formal accrediting and credentialing programmes, such as the USA, Canada and the UK it might be enough to develop measures that the approved LIS programmes in each country would accept as equivalent. However in many other parts of the world where the first professional degree is less than a Master's degree, the Guidelines would be adjusted accordingly. The various currently adopted methods of assessing quality of LIS programmes and competencies of LIS professionals were analysed and the IFLA Guidelines for Professional Library/Information Educational Programs were adopted as guiding standards. The main principles for the Quality Model are that LIS profession requires education at university level and that subject benchmarking together with diversity of LIS education programmes should be encouraged.

The Quality model includes: curriculum, learning and teaching and learning outcomes, evidencing the agreed criteria for improving mutual trust between LIS schools and fostering international collaboration for QA of LIS programmes.

1. *Curriculum*

To achieve the utmost transparency, the curriculum should be stated clearly in a publicly available document, describing the aims, prerequisites, content, learning outcomes, and assessment methods for each course within the programme. An English translation should be made available on the Web.

LIS educational programmes could be offered at technical graduate and professional level, as well as the research and doctoral level.

The LIS Schools offering the programme need to be accredited by the Government or other accrediting agency. The programme should meet such educational and/or professional accreditation requirements as are the norm in the country.

The LIS programme content should cover the knowledge areas indicated by the IFLA Guidelines. The content of a core curriculum is indicated as well, based on information management. IFLA guidelines specify theory and practice and suggest having practicum, internship and fieldwork for students. Transferable skills, such as communication skills, time management skills, analytical and problem solving skills are listed as desirable learning outcomes as well.

A process of formal curriculum review should take place on a regular basis, informed by input from employers, practitioners and professional associations, as well as from students and faculty.

2. *Learning and teaching*

Teaching and assessment methods should be designed to develop or enhance students' interpersonal communication skills, ability to work in teams, and time and task management skills. At professional level, emphasis should be placed on developing students' analytical and problem-solving skills. Instructional resources and facilities should be adequate to the minimum standard defined by the IFLA Guidelines.

The teaching staff should be sufficient to accomplish programme objectives. The qualification of each full-time faculty member should include research-based competence in the designated teaching areas, technological proficiency, effectiveness in teaching, a sustained record of scholarship, and active participation in appropriate professional associations. The educational programme should state policies and standards related to appointment, review and promotion of full-time faculty equivalent to those implemented in comparable units. All full-time faculties should hold degrees in relevant subjects from recognised academic institutions. A clearly stated policy is needed for the continuing education and professional development of the academic teaching staff, and for reviewing the currency and relevance of courses and teaching methods.

3. *Learning outcomes*

Students should be assisted in constructing a coherent programme of study to meet career aspirations consistent with the educational programme's mission, goals and objectives. Evaluation of student achievement should be provided on a consistent and equitable basis.

A clear statement of the requirements and learning objectives of the educational programme should appear in a formal document that is available to students and prospective students. Upon programme completion, students should be awarded a degree, diploma, or certificate suitable to their level of study.

A Subject benchmarking system could be established by sharing best experiences of LIS schools and to assess quality through a peer review process.

In summary, The real problem is to agree on the main actor in accreditation and recognition, as well as to agree on what to do if Professional Associations should have an active role in quality assurance. In time of change, LIS schools and the single Professional Associations should cooperate in order to obtain mutual trust and to be able to apply the principles, criteria and procedures defined by IFLA.

11.5   Conclusion

Quality criteria and indicators could act as a thinking device to promote ongoing dialogue about LIS schools quality. Regarding quality assurance of LIS education it can be argued that homogeneity does exist, despite some differences. Although an increased understanding is reported, it is of limited value in trying to achieve convergence in the formal input and process characteristics of programmes. The way programmes are organised, the delivery mode, the specific teaching and learning setting, even the exact amount of time and workload invested in them, are increasingly diverging, but this divergence does not intrinsically affect the comparability of learning outcomes. In conclusion, one could say that a quality model focused on learning outcomes can help to innovate curricular for LIS and improve quality of learning.

The content of the programme could be related to a new role of professionals in Society, which takes into consideration careful design of learning outcomes considering the core knowledge, the innovation needed and the adaptation to the local context. Learning and teaching activities should be linked to research: the research done by teachers, the research done by students, the learning of research methods by students. Learning outcomes not only should involve the students in active learning but also they should gain an international view, capability for critical thinking, self-management, professional overview and so forth, as well as familiarity with the body of knowledge of the discipline and an understanding of the social role of the profession.

The identification of appropriate learning outcomes and competencies would also facilitate the ability of employers and academic institutions to establish international reciprocity and equivalency of qualification guidelines in the global scenario of library and information professionals. The IFLA Guidelines for Equivalency and Reciprocity could be continuously updated after review and discussion by the professional community.

In conclusion, one more question could be posed in this fashion:

Does quality assurance makes a difference? The discussion is particularly important for two reasons:

- first, it prompts us to consider the need for more impact research as well as perhaps the need for a more research-informed approach to quality evaluation;
- second, it is worth reflecting on the case while the improvement of quality has been the secondary feature of most external review systems.

The main problem which has been evidenced by IFLA Special Interest Group Education and Training in Developing Countries consists in knowing how to adapt theories and concepts elaborated at international level to single situations and objectively complex contexts like libraries in developing countries (Abdullahi, 2009). Some nations are more advanced in this discipline, and have a solid academic tradition which has successfully overcome the currently ongoing changes and where the professionals are recognised by the community. Other nations, due to a less solid tradition, sometimes do not take enough advantage of the opportunities of internationalisation. These nations will benefit from participating actively in the cultural and professional debate, which has been active for years at IFLA level, by means of international associations and conferences.

References

Abdullahi, I. (Ed.). (2009). *Global library and information science: A textbook for students and educators*. Berlin: De Gruyter.

Abdullahi, I., & Kajberg, L. (2004). A study of international issues in library and information science education: Survey of LIS schools in Europe, the USA and Canada. *New Library World, 105*, 345–356.

Abdullahi, I., Kajberg, L., & Virkus, S. (2007). Internationalisation of LIS education in Europe and North America. *New Library World, 108*(1/2), 7–24.

Audunson, R., Nordlie, R., & Spangen, I. C. (2003). The complete librarian – An outdated species? LIS between profession and discipline. *New Library World, 104*(1189), 195–202.

Boaz, M. (1986). International education: An imperative need. *Journal of Education for Library and Information Science, 3*(Winter), 165–170.

Brine, A., & Feather, J. (2003). Building a skills portfolio for the information professionals. *New Library World, 104*(1194/1195), 455–463.

Burnett, K., & Bonnici, L. (2006). Contested terrain: Accreditation and the future of the profession of librarianship. *Library Quarterly, 76*(2), 193–219.

Campbell, C., & Van der Wende, M. C. (2000). *International initiatives and trends in quality assurance for European Higher Education*. Helsinki: ENQA.

Chu, H. (2010). Library and information science education in the digital age. In A. Woodsworth (Ed.), *Exploring the digital frontier* (pp. 77–112). London: Emerald Group Publishing.

Dalton, P., & Levinson, K. (2000). An investigation of LIS qualifications throughout the world. *66th IFLA Council and General Conference in Jerusalem, 13–18 August*. Retrieved September, 2013, from http://archive.ifla.org/IV/ifla66/papers/061-161e.htm.

ENIC-NARIC net. (2008). *Gateway to recognition of academic and professional qualifications*. [Online]. Retrieved September, 2013, from http://www.enic-naric.net/.

Fang, G. R., Nauta, P., & Fischer, E. (1987). Guidelines to equivalence and reciprocity of professional qualifications. *IFLA Journal, 13*(2), 133–140.

Harvey, L. (1995). Beyond TQM. *Quality in Higher Education, 1*(2), 123–146.

Huckle, M. (2002). Driving change in the profession: Subject benchmarking in UK library and information management. *Libri, 52*, 209–213.

IFLA. Education and Training Section. (2012). *Guidelines for LIS educational programmes; updated version, by Smith, K., Hallan, G., Ghosh*. Retrieved September, 2013, from http://www.ifla.org/publications/guidelines-for-professional-libraryinformation-educational-programs-2012.

IFLA. Section Education and Training. (2007). *World guide to library and information science education, updated version, by Schniederjürgen A*. Munich: K.G. Saur.

IFLA. Section Education and Training. (2008). *Feasibility of international guidelines for equivalency and reciprocity of qualifications for LIS professionals, by Tammaro, A. M. & Weech, T*. Retrieved September, 2013, from http://www.ifla.org/publications/feasibility-of-international-guidelines-for-equivalency-and-reciprocity-of-qualificatio.

IFLA. Section Education and Training. (2009). *International guidelines for equivalency and reciprocity of qualifications for LIS professionals. Draft guidance document for transparency, equivalency and recognition of qualifications by Terry Weech, Anna Maria Tammaro.* Retrieved September, 2013, from http://www.ifla.org/publications/international-guidelines-for-equivalency-and-reciprocity-of-qualifications-for-lis-prof.

Kajberg, L. (2002). Cross-country partnerships in European library and information science: Education at the crossroads. *Library Review, 51*, 164–170.

Kajberg, L., & Lorring, L. (Eds.). (2005). *European curriculum reflections on library and information science education.* Copenhagen: Royal School of Library and Information Science. [Online]. Retrieved September, 2013, from http://dspace-unipr.cilea.it/handle/1889/1704.

Khoo, C., Majid, S., & Sattar Chaudry, A. (2003). Developing an accreditation system for LIS professional education programmes in Southeast Asia: Issues and perspectives. *Malaysian Journal of Library and Information Science, 8*, 131–149.

Knox, A. B. (2001). Strengthening the quality of continuing professional education: Delivering lifelong continuing professional education across space and time. In C. H. Vermount (Ed.), *4th World conference on continuing professional education for the Library and Information Science professionals.* SAUR: Munchen.

Myburg, S., & Tammaro, A. M. (2013). *Exploring education for digital librarians: Meaning, modes and models.* London: Chandos.

OECD. (1996). *Internationalisation of higher education.* Paris: OECD.

Tammaro, A. M. (2005). Report on quality assurance models in LIS programs. IFLA Education and training Section. Retrieved September 2013 from http://www.ifla.org/files/assets/set/s23_Report-QA-2005.pdf.

Tammaro, A. M. (2006). Quality assurance in library and information science (LIS) schools: Major trends and issues. In D. A. Nitecki & E. G. Abels (Eds.), *Advances in librarianship* (Vol. 3). London: Emerald.

Tammaro, A. M. (2007). *Performance indicators in library and information science (LIS) education: Towards cross-border quality assurance in Europe.* Retrieved September, 2013, from www.cbpq.qc.ca/congres/congres2007/Actes/Tammaro.pdf.

Tammaro, A. M., & Dixon, P. (2003). Strengths and issues in implementing a collaborative inter-university course: The international masters in information studies by distance. *Education for Information, 21*(2/3), 12–27.

Tuning. (2004). Tuning educational structure in Europe; closing conference document. Tuning Project 2. Retrieved September, 2013, from http://www.aic.lv/bolona/Bologna/Reports/projects/Tuning/Tun_Book.pdf.

UNESCO. (1984). *International symposium on the harmonisation of education and training programmes in information science, librarianship and archival studies. Final report.* Paris: UNESCO. October 8–12.

UNESCO. (2002). *First global forum on quality assurance, accreditation and the recognition of qualifications.* Paris: UNESCO.

UNESCO. (2004). *Second global forum on international quality assurance, accreditation and recognition of quality assurance: Widening access to quality higher education.* Paris: UNESCO.

UNESCO CEPES. (2007). *Quality assurance and accreditation: A glossary of basic terms and definitions by Vlasceanu L, Grunberg L, Parlea, D.* UNESCO CEPES: Bucharest.

UNESCO International and Information Unit. (1998). *A curriculum for an information society.* Bangkok: UNESCO.

Van House, N., & Sutton, S. A. (1996). The Panda syndrome: An ecology of LIS education. *Journal of Education for Library and Information Science, 37*, 131–147.

Virkus, S., & Hartley, R. J. (2003). Approaches to quality assurance and accreditation of LIS programmes: Experiences from Estonia and United Kingdom. *Education for Information, 21*, 31–48.

Wozniczka-Paruzel, B. (2003). Experiences of library and information science (LIS) studies accreditation in the context of quality assurance systems in Poland. *Education for Information, 21*, 49–57.

WTO. (1998). *Education services. Background note by the secretariat.* Geneva: WTO.

Chapter 12
Accreditation in North America, A unique Quality Assurance Program

Beverly P. Lynch

12.1 Background on Accreditation in the USA

Accreditation of institutions of higher education in the USA began in the early 1900s in efforts to define what was a high school? A college? A Medical school? These efforts sought to address matters relating to the issues of articulation between high schools and colleges. Over the years accreditation has changed from a quantitative approach with specific requirements to a qualitative approach based on more general standards, and from an emphasis on making institutions more alike to recognizing and encouraging institutional individuality. Initially the system of accreditation was heavily dependent on external reviews. Now it is based more on self-evaluation and self-regulation. At the outset the focus was on judging an institution; now its primary goal is that of encouraging an institution to improve its educational quality.

In the USA there is a great diversity of institutions, and programs. Given the diversity in over 4,300 institutions in the USA, it is clearly impossible to establish standardized requirements and administer them through some national agency. The magnitude of this diversity has led to the development of accreditation as a voluntary, nongovernmental process of self-regulation focused on evaluating and improving educational quality. The basic characteristics of accreditation that have emerged are (1) prevailing sense of voluntarism, (2) a strong tradition of self-regulation, (3) a reliance on evaluation techniques, and (4) its primary concern with quality (Young, Charles, Kells, & Associates, 1983, 11). An agreed upon definition is

> a process by which an institution of post-secondary education evaluates its educational activities, in whole or in part, and seeks an independent judgment to confirm that it is substantially achieving its objectives and is generally equal in quality to comparable institutions or specialized units. (Young et al., 1983, xi).

B.P. Lynch (✉)
University of California, Los Angeles, CA, USA
e-mail: lynch@gseis.ucla.edu

M. Miwa and S. Miyahara (eds.), *Quality Assurance in LIS Education:*
An International and Comparative Study, DOI 10.1007/978-1-4614-6495-2_12,
© Springer Science+Business Media New York 2015

In the USA accreditation serves as a process of evaluation and not of regulation. There is a strong consensus that accreditation in the USA is essentially nongovernmental, and that it is centered on a voluntary system of self-regulation. In these elements accreditation is unique to the USA. Other nations place the official responsibility for establishing and maintaining educational standards in a government agency.

The major characteristics of accreditation in the USA are:

1. It is predominantly a voluntary, private-sector activity and therefore cannot mandate compliance or control behavior except by persuasion and peer influence.
2. It is the premier example of self-regulation (as opposed to government regulation in postsecondary education).
3. It focuses primarily on judging educational quality—an elusive concept- and, given the great diversity of postsecondary educational institutions in the USA, criteria tend to be general and variable.
4. It functions essentially as an evaluative process, and institutional self-study is at the heart of the process.
5. It provides outside consultation, closely tied to the institution's own research and planning (Young et al., 1983, 21–23).

What accreditation is not is that it is

1. Not governmental although federal and state agencies use it;
2. Not mandatory;
3. Not a rating system;
4. Not a stamp of approval on individual students or courses, although it has been perceived that way.

12.2 Institutional and Specialized Accreditation

Institutional accrediting bodies help an institution look at itself as a whole. Specialized accrediting bodies such as the American Library Association (ALA), which accredits programs in library and information science for the USA and Canada, focus mainly on programs that are placed within the institution and generally have specific standards relating to skills considered desirable by practitioners as well as by standards of good practice determined by consensus. Most specialized accrediting bodies accredit programs that are units such as departments or schools within institutions. Often the agencies that accredit these specialized programs are units within institutions such as the Committee on Accreditation within the American Library Association. These professional organizations have many activities, and accreditation can be only one of these. In large, well established professional organizations, accreditation, while considered important, is not located high in the organizational hierarchy. In newly established specializations, however, accreditation is seen as a major force for achieving recognition and acceptance.

Specialized accreditation and its processes and procedures are influenced by institutional accreditation practices and in many instances the specialized accreditation can occur only if the program being evaluated is organized as a unit within an institution that is accredited by one of the six regional accrediting organizations in the USA. This requirement has been relaxed somewhat depending upon particular circumstances.

The sections that follow describe the accreditation of library and information science programs in the USA and Canada which is carried out by the Committee on Accreditation of the American Library Association. It acknowledges that individuals with degrees from other countries can be considered as having accredited degrees, but these are not common and require applications and considerations of the other countries' degree.

12.3 Early Efforts at Quality Control in Library Science in the USA

Melvil Dewey's School, founded in 1887, was the first systematic program of library education in the USA. There were, however, earlier calls for education for librarianship and American librarians were aware of the developments in Germany, particularly a proposal authored by Dr. F. Rullmann, which referred to a paper published by Schrettinger in 1834 who had advocated for a special school of education for librarians (Lynch, 2008, 934). Dewey proceeded without references to Rullmann or Schrettinger. Many have criticized Dewey for not setting the right direction for education in librarianship in the early years, but there were no methods in place to evaluate the program. Dewey, unlike Rullmann and Schrettinger, was less interested in librarianship as it might be practiced in university libraries. His interest, although never really articulated, was on the need for education and training for personnel in the growing numbers of public libraries and in the small college libraries that continued to be established in the USA.

In 1915, the Carnegie Corporation authorized an inquiry into library schools and the adequacy of the training of librarians. The Corporation had provided support for the establishment of four library training programs, one in a university, Western Reserve University, and three in public libraries: the Carnegie Library of Pittsburgh, the Carnegie Library of Atlanta and the New York Public Library. The Corporation also was receiving many requests for grants to establish more training programs and to fund the building of local public libraries. The report to the Carnegie Corporation by Alvin S. Johnson in 1916 provided some important observations on library training. While his recommendations were not implemented, his report provided the impetus for the Corporation to undertake the task of assessing library education and making recommendations for its future. It gave this assignment to C.C. Williamson whose influential reports, sponsored by the Carnegie Corporation, were published in 1921 and 1923. The Williamson reports have guided the development of library education in the USA since that time (Vann, 1971).

Williamson was scathing in his assessment of the programs he reviewed. He recommended that professional training be based on a thorough college course of four years followed by at least one year of graduate study in a properly organized library school (Williamson, 1923, 136). This recommendation was adopted by the profession over the years and continues to this day. In 1951 the American Library Association established the policy that the master's degree is the professional degree in the field. Williamson commented on the curriculum, the methods of instruction, the credentials (or lack of them) of the faculty. A major recommendation was that "...the professional library school should be organized as a department of a university, along with other professional schools, rather than in public libraries, state or municipal." (Williamson, 1923, 142). This recommendation continued to guide the profession and reflects the development within American higher education that professional programs of study be carried out within the context of the university. A study published in 1979 compared the educational requirements for librarianship with medicine, law, social work, teaching, and nursing and found that "librarianship has exhibited an evolution that parallels, and at times even anticipates, the other professions" (Heim, 1979, 131). The library school programs located at the time of the Williamson reports within the various public libraries either closed or were merged into local universities.

While Williamson was publishing his report, the American Library Association was establishing its ALA Board of Education for Librarianship with the purpose of developing minimum standards for library education programs. The American Library Association has continued its interest in developing standards for library education programs and in implementing a program for accreditation of these programs. The Association has routinely reviewed and revised the standards for the accreditation of library programs, beginning with the first efforts of the ALA Board of Education for Librarianship which issued standards in 1925 and 1926. These early standards were generally quantitative in nature. Subsequent revisions of the educational standards appeared in 1933, 1951, 1972, 1992, 2008, and over the years the standards have become less quantitative and more qualitative in nature and reflect the subsequent development and change in accreditation in general.

Williamson considered the matter of certification of librarians and standards for library schools. He acknowledged that librarianship is similar to a group of professions like architecture and engineering where the first step in developing standards is best taken through voluntary action by the profession itself (Williamson, 1923, 144) and he acknowledged the potential role of the American Library Association in this effort.

More professional organizations in library and information studies have emerged in the information field since the early days of accreditation and many of them, seeking recognition and approval for the particular specialization they emphasize, turn to accreditation as a method of recognition. A cooperative study carried out in 1985 brought together 17 organizations, including the American Library Association, to discuss whether the accreditation process should be broadened in its governance and operations to include these organizations. The organizations ultimately agreed that the ALA should continue to manage the accreditation process and standards development for the field of library and information science (American Library Association, 1986). The process and the standards now in place reflect the current policies and practices related to accreditation in general: a qualitative approach based on more general standards, a focus on recognizing and encouraging institutional individuality, and a process based on self-evaluation and self-regulation. The primary goal of specialized accreditation, as is that of general accreditation, is that of encouraging a program to improve its educational quality.

12.4   The Process for Accreditation of Master's Programs in Library and Information Studies

Beginning in the 1980s library schools recognized that the scope of responsibility for education in the field was expanding. Information professionals closely related to librarianship—archivists, records managers, indexers, abstractors, database service specialists, and others—were not comfortable with the educational programs of library schools, and students were finding jobs in these related fields, so the schools expanded their curricula and changed their names by adding "information science" or "information studies." They added to the educational base of the faculty with other specializations so as to acknowledge the growing complexity of the field. Accreditation of the field continued through the program managed by the Committee on Accreditation (COA) and the Office for Accreditation of the American Library Association. It should be pointed out that there are several types of educational programs in the field of librarianship that are not evaluated through the ALA accreditation program. These include

- Doctoral programs offered by schools of library and information science are not covered by accreditation;
- Library technician programs that focus on undergraduate, paraprofessional training are not covered by ALA accreditation;
- School librarian programs that are outside the purview of the ALA accreditation process, particularly those offering a specialization in school librarianship from an educational unit accredited by the National Council for the Accreditation of Teacher Education (NCATE) and recognized by the American Association of School Librarians (AASL) a division of the American Library Association. These programs usually are associated with colleges or schools of education, that prepare students for state certification as teachers and school library media specialists

Currently there are 63 programs in library and information science accredited by the ALA; 6 of these are in Canadian universities. The ALA is one of 20 members of the Association of Specialized and Professional Accreditors (ASPA) that accredit educational programs in both the USA and Canada.

The ALA standards define the field of library and information studies as being "…concerned with recordable information and knowledge and the services and technologies to facility their management and use." The standards state further that the field

encompasses information and knowledge creation, communication, identification, selection, acquisition, organization and description, storage and retrieval, preservation, analysis, interpretation, evaluation, synthesis, dissemination, and management (Standards, 2008, 3.)

The requirements in the standards apply regardless of the forms of the locations of delivery of a program. The definition incorporates a field of professional practice and associated areas of study and research. "School of library and information studies" means that unit organized and maintained by an institution of higher education for the purpose of graduate education in library and information studies (Standards, 2008, 3).

Central to the implementation of the standards in each program is a systematic planning process which requires a continuous review and revision of the program's vision, mission, goals, objectives and learning outcomes, and assessment of the attainment of goals and objectives. Planning efforts to redesign core activities and communication of the planning policies and processes also are required. These broad-based systematic planning efforts require the involvement of program constituents and a thorough and open documentation of those activities required in planning (Standards, 2008, 4).

One of the primary functions of the Office of Accreditation of the American Library Association is to administer the review process. The review process requires the preparation of the self-study, the "Program Presentation" document, at least every 7 years as part of the comprehensive review process, an annual statistical report, and a biennial report that addresses changes in the program since the last comprehensive review. These activities are described in the *Accreditation Process Policies and Procedures* prepared and distributed by the ALA Office for Accreditation (Committee on Accreditation of the American Library Association 2012). The document also provides a detailed description of the process, and overview of the American Library Association's definitions and terminology pertaining to accreditation.

The Committee on Accreditation of the ALA is the agency doing the accreditation. Its work is supported by the ALA Office for Accreditation. The committee has 12 members appointed by the President of the ALA for a one four year term which is not renewable. Two of the 12 members are not affiliated with the field of library and information studies in any way, they are appointed for 2 year terms, renewable once, to represent the public interest. The purposes of accreditation, as implemented by the Committee on Accreditation, include accreditation of only the first professional degree programs. It does not accredit doctoral programs, undergraduate programs, certificate programs or continuing education programs. The Committee accredits programs, not schools or institutions, and in accrediting programs it does not certify individuals. A site visit is the means by which the Committee on Accreditation reviews those aspects of a program that cannot be evaluated by documentation alone. The visiting team is appointed by the COA following the comments of the program as to suitability of the members. The team follows the procedures for the site visit as established by the COA and concludes the visit with a report that includes a factual section, an evaluative section, and a set of recommendations for improvement of the program. The draft of the factual section is mailed to the program within 10 days of the site visit for verification and correction. Since the evaluative section must be based on the factual section, this is an important step in the process. The response from the program might lead to correction of the factual section and possible revision of the evaluative section and recommendations. The final site visit report as a whole is sent to the COA which forwards the report except for a final recommendation concerning the accreditation decision, to the school which then has an opportunity to respond to it in writing or in person. All of these reports, as is the entire process, are treated confidentially by the COA and the site-team members. However, the program can make known the content of the final report as it wishes.

12.5 The Standards for Accreditation of Master's Programs in Library and Information Studies Adopted by the Council of the American Library Association January 15, 2008

The current standards were adopted by the Council of the American Library Association in 2008. The Committee routinely reviews the standards as its regular meetings. The standards also are under review by the profession itself to determine whether they remain applicable to an ever changing profession. As one can see over the years, revision is not a quick process, a normal time period for revision can be upwards of 20 years.

The standards specify six areas for determining program quality: Standard I: Mission, Goals, and Objectives; Standard II: Curriculum; Standard III: Faculty; Standard IV: Students; Standard V: Administration and Financial Support; and Standard VI: Physical Resources and Facilities. As was mentioned above the implementation of a board-based systematic planning process that includes the constituency that a program seeks to serve is central to the implementation of the standards.

12.5.1 Standard I: Mission, Goals, and Objectives

On-going planning is required in the first statement in this standard. It states that the program goals of the school and its objectives are to foster quality education. The second statement requires the program objectives to be stated in terms of learning outcomes. The objectives are to reflect: "the essential character of the field of library and information studies; that is, recordable information and knowledge, and the services and technologies to facilitate their management and use." These are to encompass information and knowledge creation, communication, identification, selection, acquisition, organization and description, storage and retrieval, preservation, analysis, interpretation, evaluation, synthesis, dissemination, and management The first standard refers to philosophy, principles, and ethics of the field, the value of teaching and service and the important of research, and the needs of the constituencies the program seeks to serve. Standard I emphasizes that each program is judged on the degree to which it attains its own objectives. So within the context of the mission of the school, program goals and objectives form the essential frame of reference for meaningful external and internal evaluation (Standards, 6–7). That standards do not prescribe a school's mission, goals, and objectives, the school does that. Within the context of accreditation, the program is evaluated on the degree to which it attains its objectives. Standard I, like the standards which follow, emphasize that the evaluation of the program goals and objectives involves those served: students, faculty, employers, alumni, and other constituencies.

12.5.2 Standard II: Curriculum

The first statement in this standard also emphasizes the importance of program's goals and objectives and its systematic ongoing planning process. The standard defines the curriculum:

> The curriculum is concerned with recordable information and knowledge, and the services and technologies to facilitate their management and use... The curriculum of library and information studies encompasses information and knowledge creation, communication, identification, selection, acquisition, organization and description, storage and retrieval, preservation, analysis, interpretation, evaluation, synthesis, dissemination and management. (Standards 7)

The standard comments on the study of services and activities in specialized fields, building these experiences upon a general foundation of library and information studies. The statements of knowledge and competencies developed by relevant professional organizations are to be taken into account when a program includes these learning experiences in the curriculum.

12.5.3 Standard III: Faculty

There is an emphasis on the full-time faculty being qualified for appointment to the graduate faculty of the parent institution and being of sufficient number with a diversity of specializations necessary to carryout the major share of teaching, research, and service in the program. Part-time faculty are acknowledged and the standard comments that they are expected to balance and complement the full-time faculty. Full-time faculty are expected to have a sustained record of accomplishment in research and that procedures are in place to enable a systematic evaluation of faculty in the areas of teaching, research, and service.

12.5.4   Standard IV: Students

The standard regarding students indicates that the policies regarding recruitment, admission, financial aid, placement, are consistent with the program's goals and objectives and its mission. It also requires the school to have recruitment and retention policies that support the diversity of North America's communities. The admission standards are to be applied consistently and students admitted to a master's program in library and information studies are to have a bachelor's degree from an accredited institution. Students are expected to construct coherent programs of study, to receive evaluation of their achievements, and to have access to opportunities for guidance, counseling, and placement. Participation in student organizations and in determination of policies relating to academic and student affairs also is available to students.

12.5.5   Standard V: Administration and Financial Support

The standard states that the school is an integral yet distinctive academic unit within the institution with autonomy sufficient to assure the attainment of the program's objectives. The parent institution is expected to provide the resources and administrative support necessary to attain the goals and objectives. The emphasis is that the level of support provided by the parent institution is sufficient to develop and maintain library and information studies education in accordance with the general principles set forth in the standards.

12.5.6   Standard VI: Physical Resources and Facilities

The program is to have access to physical resources and facilities, including access to library and multimedia resources and services, computer and other information technologies, accommodations for independent study and media production facilities, that are sufficient to attainment of the program's goals.

An afterword to the standards describes the process used in the review and revision. It also states that the "The unit called a "school" may be organized as an autonomous college within its university, as a department in a college, or otherwise, as appropriate within its institution." Further it reiterates that programs are evaluated in the same way regardless of locations or forms of delivery of a program (Standards, 13).

12.6   Conclusion

Accreditation has been an integral part of education for library and information studies in North America for nearly 100 years. The standards used in accrediting the educational programs have been reviewed and revised regularly and, as a result, accreditation of the educational programs has reflected the changes in accreditation generally, and changes in the field itself. The Standards and the process of implementing the standards give a unique role to the profession as it is practiced and to the profession as it is taught. While the responsibility for the process central to the accreditation activities is placed in the American Library Association, the many other organizations in the information field also have a place and a role in the process. This is acknowledged by the many parent institutions that support library and information science education, as an important aspect to support the parent institutions give to these educational programs.

The nongovernmental aspect to accreditation as practiced in the USA, the well-developed review process, and the clarity and brevity of the standards have been well received by the profession. The voluntary and self-regulatory nature of accreditation in the USA, may provide a model to meet accreditation requirements on a more global scale as global demand continues to grow and educational programs develop to meet it.

References

Committee on Accreditation of the American Library Association. (2012). Accreditation Process, Policies, and Procedures (AP3), Third Edition, Chicago: American Library Association, 68p.

American Library Association. (1986). *Accreditation: A way ahead*. American Library Association.

Heim, K. M. (1979). Professional education: Some comparisons. In *As much to learn as to teach: Essays in honor of Lester Asheim*. Hamden, CT: Linnet.

Lynch, B. P. (2008). Library education: Its past, its present, its future. *Library Trends, 56*(4), 931–953.

(2008). *Standards for Accreditation of Master's Programs in Library & Information Studies*: Adopted by the Council of the American Library Association January 15, 2008. Chicago: American Library Association.

Vann, S. M. (1971). *The Williamson reports: A study*. Metuchen, NJ: Scarecrow Press.

Williamson, C. C. (1923). Training for library service: A report prepared for the Carnegie Corporation of New York. New York.

Young, K. E., Charles, M. C., Kells, H. R., & Associates. (1983). *Understanding accreditation*. San Francisco: Jossey-Bass.

Chapter 13
Accreditation Processes in Latin America: An Exploration into the Cases of Library and Information Science (LIS) Programs in Mexico, Colombia, and Costa Rica

Mónica Arakaki

13.1 Introduction

Over the 2000–2010 decade, Latin America has consolidated a higher education model that started to take form during the 1980s, immersed in a dynamic fueled by the expansion of the national economies and the consequent demand for qualified workers (Rama 2011, 16). Apart from the dramatic increase of student enrollment in tertiary education—from approximately 11 million in 2000 to 22 million in 2010, especially at private institutions—the region has been characterized by growing concerns about quality and internationalization, coming from both higher education institutions (HEIs) and national authorities. Regarded fundamentally as a public good, education has been subject to increasing government regulations during this 10-year period. As Rama (2011, 16) indicates, public policies have favored "more standards and scrutinizing procedures for authorizing HEIs to start operations (in almost all the region), accountability systems (Colombia and Mexico), budget negotiations (Costa Rica), specific incentives (Mexico), creation of evaluation and accreditation agencies (Chile, Venezuela, Panama, Nicaragua, Brazil, Bolivia, Peru, etc.), and procedures for professional certification (Brazil, Mexico, and Colombia)."

Latin America was among the first regions in the world to embrace quality assurance systems in education. By the end of the 1990s, Chile, Colombia, and Argentina had already established national systems (Lemaitre & Zenteno, 2012, 26) and since then, most countries have implemented mechanisms for publicly acknowledging the quality of the education delivered. "Each country had been finding its path and exploring ways to better evaluate and accredit institutions and programs, and align them to its own needs, history and the particular characteristics of its government regulations and tertiary education system" (Brunner & Ferrada Hurtado, 2011, 401). However, not all have reached the same level of maturity in this regard. Whereas

M. Arakaki (✉)
Pontifical Catholic University of Peru, Lima, Peru
e-mail: monica.arakaki@pucp.pe

M. Miwa and S. Miyahara (eds.), *Quality Assurance in LIS Education:*
An International and Comparative Study, DOI 10.1007/978-1-4614-6495-2_13,
© Springer Science+Business Media New York 2015

Argentina, Colombia, Chile, Costa Rica, and Mexico exhibit relatively well-developed systems; Brazil, Ecuador, Panama, Paraguay, Peru, and Uruguay are at an earlier stage of development; and agencies in Bolivia and Venezuela have been recently formed or are in the process of being formed (Lemaitre & Zenteno, 2012, 34–38).

It is in this context that Library and Information Science (LIS) education takes place. LIS programs are delivered by public and private HEIs in almost all the countries of the region, prominently at undergraduate level (2–3 years for associate's degrees and 4–6 years for bachelor's/licentiate degrees); doctoral degrees are awarded only in Brazil and Mexico. Unfortunately the literature about quality assurance in LIS education remains scarce. Although there are important regional and national initiatives, projects, and studies about quality assurance in higher education, documented experiences in the LIS field could hardly be found. Among these, Escalona Ríos (2011) and Múnera Torres (2012) present an overview of quality assurance measures taken by Latin American LIS schools. As pointed out, many schools have applied some sort of review, as required by institutional standard procedures or promoted by grassroots initiatives. However, only in Mexico, Colombia, and Costa Rica, LIS schools were found to have committed themselves to getting their programs accredited by national agencies. In the particular case of Colombia, the HEIs where LIS schools belong gained institutional accreditation as well. This article takes an exploratory approach to describe the accreditation processes undertaken by this group of LIS schools, as encouraged by their respective national quality assurance systems and according to the regulations, frameworks, procedures, and models they follow.

13.2 Mexico

The Mexican quality assurance system in higher education—the National System of Evaluation, Accreditation, and Certification—is comprised of several governmental and nongovernmental organizations, each one with different tasks (evaluation, assessment, accreditation, certification) and scopes (students, faculty, academic programs, accreditation agencies). Since the 1970s, and especially since the 1990s, these organizations have been developing a set of reference frameworks, criteria, indicators, performance measures, standards, instruments, and promotion strategies which seek to contribute to the continuous improvement of higher education, as a means to attain social equity (Rubio Oca, 2007, 36).

As can be seen from Table 13.1, Higher education institutions (universities, institutes of technology, technological universities, polytechnic universities, colleges, or normal schools) are directly responsible for assessing their students, and evaluating their faculty and programs of study, but they may also be subject to external evaluations.

13.2.1 Minimum Requirements

The voluntary affiliation to the National Association of Universities and Higher Education Institutions (ANUIES) constitutes a mechanism of self-regulation among higher education institutions (HEIs) and one first step towards quality education.

Table 13.1 National System of Evaluation, Accreditation, and Certification (Mexico)

Scope	Organization	
Students	HEI	Higher Education Institution
	CENEVAL	National Center for Evaluation in Higher Education
Faculty	HEI	Higher Education Institution
	SNI	National System of Researchers
Undergraduate programs	HEI	Higher Education Institution
	CIEES	Inter-institutional Committees for Evaluation in Higher Education
Graduate programs	HEI	Higher Education Institution
	CIEES	Inter-institutional Committees for Evaluation in Higher Education
	SEP and	Ministry of Public Education
	CONACYT	National Council for Science and Technology
Accreditation agencies	COPAES	Council for the Accreditation of Higher Education

ANUIES promotes "the comprehensive and continuous improvement of the programs and services provided by its associates, and the whole higher education system" (ANUIES, 2006). Established in 1950, this association currently groups 175 public and private HEIs, which concentrate 80 % of Mexico's student enrollment in undergraduate and graduate programs (ANUIES, 2012). For the HEIs, the procedure to join ANUIES involves an evaluation in terms of institutional regulations, planning, infrastructure, curricula, internal evaluation procedures, faculty, students, financial resources, and cultural activities (Fresán Orozco & Taborga Torrico, 2002, 26).

After joining ANUIES, HEIs may voluntarily take further steps: submit themselves to a diagnostic evaluation and if the results are favorable, request an evaluation for accreditation purposes. These are two different—although complementary—processes, the first being a requirement for the second. Whereas the first is aimed at identifying aspects and lines of action for improvement, the second seeks to determine how well an academic program or unit complies with a set of predefined standards.

13.2.2 Diagnostic Evaluation

Evaluations of academic programs, services and projects are conducted by nongovernmental organizations called Inter-institutional Committees for Evaluation in Higher Education (CIEES). Created in 1991 as an agreement between the Ministry of Public Education (SEP) and ANUIES, the CIEES have regarded academics among their members, which are organized in committees according to their area of expertise. There are nine committees: Administration and Institutional Management; Architecture, Design and Urban Planning; Arts, Education and Humanities; Agricultural Science; Health Sciences; Natural and Exact Sciences; Social Sciences and Management; Cultural Affairs; and Engineering and Technology (CIEES, 2001).

Table 13.2 Number of programs evaluated by the CIEES (Mexico)

	Number of programs	Percentage of programs
Class 1	2,802	66.4 %
Class 2	1,128	26.7 %
Class 3	290	6.9 %
Total	4,220	100 %

Table 13.3 Class 1 LIS undergraduate programs (Mexico)

HEI	Academic program	Year of award
Autonomous University of Chiapas (UNACH)	Library Science and Information Management	2008
Autonomous University of Chihuahua (UACH)	Information Science	2008
Autonomous University of Mexico State (UAMex)	Documentary Information Sciences	2006
Autonomous University of Nuevo Leon (UANL)	Library and Information Science	2002
Autonomous University of San Luis Potosi (UASLP)	Library and Information Studies	2004
National Autonomous University of Mexico (UNAM)	Library and Information Studies	2007

According to the CIEES, these evaluations are "continuous, comprehensive, and participatory processes that allow for the definition, analysis, and explanation of a problematic situation through relevant information" (ANUIES, 1997). Based on previous self-evaluations, they take into account several aspects (faculty, students, curriculum, assessment of learning, learning support services, research, infrastructure, equipment, and management and funding) according to frameworks recognized by the corresponding scholar community.

After pondering all the elements involved, the academic peers place the program in a category depending on the degree of achievement attained:

- Class 1: for programs with high degree of accomplishment, with the possibility of getting accredited in the short term.
- Class 2: for programs with intermediate degree of accomplishment, with the possibility of getting accredited in the medium term.
- Class 3: for programs with low degree of accomplishment, with the possibility of getting accredited in the long term.

As of January 2013, the CIEES have evaluated a total of 4,220 undergraduate and graduate programs, 66.8 % of which obtained Class 1 category (CIEES, 2013), as shown by Table 13.2.

As the rest of higher education programs in Mexico, LIS studies are offered by seven HEIs at three levels: (1) undergraduate programs that award associate degrees after 2 years of study, (2) undergraduate programs that award licentiate degrees (bachelor's degrees) after 4 or 5 years of study, and (3) graduate programs that award master's or doctoral degrees. A list of programs, HEIs, and academic departments is provided in Appendix 1.

To date, six LIS programs have been granted Class 1 category, all at the undergraduate level (CIEES, 2013) (Table 13.3).

13.2.3 Evaluation with Accreditation Purposes: Undergraduate Programs

Higher education programs are accredited by a number of specialized nongovernmental agencies. These accreditation agencies are authorized and supervised by the Council for the Accreditation of Higher Education (COPAES), an association created in 2000 and officially acknowledged by SEP. COPAES seeks to guarantee that all accreditation agencies follow academically rigorous, unbiased, and transparent procedures. At the present time, COPAES has authorized 28 accreditation agencies to operate (COPAES, 2013). These agencies, whose members are recognized scholars and professionals, are specialized in one area of knowledge.

The category of accredited (valid for 5 years) is conferred to those academic programs deemed to be "socially relevant," coherent with their respective institutional plans, and equipped to meet their strategic goals. These attributes are operationally defined using a set of criteria and measured through indicators and standards. Actors, processes, and outcomes involved are examined against 49 criteria, organized in ten categories (COPAES, 2012) (Table 13.4):

Table 13.4 COPAES' criteria for accreditation (Mexico)

Category	Criterion
Faculty	Recruitment
	Selection
	Hiring
	Development
	Full-time and adjunct faculty
	Distribution of academic hours for full-time faculty
	Assessment
	Promotion
Students	Selection
	Admission
	Track records
	Group size
	Graduation
	Academic performance measures
Curriculum	Education model
	Entry and exit profiles
	Regulations
	Syllabi
	Contents
	Flexibility
	Evaluation and update
	Publicity
Assessment of learning	Methodology
	Achievement encouragement

(continued)

Table 13.4 (continued)

Category	Criterion
Integral education	Entrepreneurship development
	Cultural activities
	Sports activities
	Career counseling
	Risk attitudes prevention counseling
	Health services
	Family involvement
Learning support services	Institutional tutorship program
	Academic counseling
	Library
Extension services	Connections with public, private and social sectors
	Alumni track records
	Academic exchange
	Service learning
	Job vacancies service
	Extension courses
Research	Research lines and projects
	Resources
	Dissemination
	Impact
Infrastructure and equipment	Infrastructure and facilities
	Equipment
Management and funding	Planning, evaluation, and organization
	Administrative staff
	Financial resources
Total	49

As mentioned before, diagnostic evaluations and evaluations for accreditation purposes are two distinct, but closely related processes. While the first are "complex, thorough, analytical, and explanatory," being expected to diagnose problems and provide suggestions for improvement, the second, which measure parameters and compare them against standards, involve a relatively more simple task (ANUIES, 1997). Nevertheless, both are based on the same reference framework of categories and criteria, in order to maintain a consistent evaluation system (Rubio Oca, 2007, 40). In fact, Class 1 status is equivalent to a temporary accreditation for those academic programs in disciplines that are not covered by any of the accreditation agencies recognized by COPAES (Martínez Rider, 2010, 242).

The accreditation agencies relevant to LIS programs are the Association for the Accreditation and Certification in Social Sciences (ACCECISO) and the Council for the Accreditation of Academic Programs in Humanities (COAPEHUM). LIS programs, offered by schools that mostly belong under faculties of philosophy and humanities, were accredited in the first years by ACCECISO (created in 2002), considering social sciences as the closest area of knowledge. More recently, COAPEHUM (created in 2007) has been assigned this responsibility. Up to 2012, ACCECISO and

Table 13.5 Accredited LIS undergraduate programs (Mexico)

HEI	Academic program	Year of award	Accreditation agency
Autonomous University of Chiapas (UNACH)	Library Science and Information Management	2010	ACCECISO
Autonomous University of Mexico State (UAMex)	Documentary Information Sciences	2007	ACCECISO
Autonomous University of Nuevo Leon (UANL)	Library and Information Science	2009	ACCECISO
Autonomous University of San Luis Potosi (UASLP)	Library and Information Studies	2009	ACCECISO
National Autonomous University of Mexico (UNAM)	Library and Information Studies	2012	COAPEHUM

Table 13.6 Class 1 and accredited LIS programs (Mexico)

HEI	Program	Class 1 CIEES	Accredited ACCECISO	COAPEHUM
Autonomous University of Chiapas (UNACH)	Library Science and Information Management	2008	2010	
Autonomous University of Chihuahua (UACH)	Information Science	2008		
Autonomous University of Mexico State (UAEMex)	Documentary Information Sciences	2006	2007	In progress
Autonomous University of Nuevo Leon (UANL)	Library and Information Science	2002	2009	
Autonomous University of San Luis Potosi (UASLP)	Library and Information Studies	2004	2009	
National Autonomous University of Mexico (UNAM)	Library and Information Studies	2007		2012

COAPEHUM have accredited 247 and 64 programs respectively (ACCECISO, 2012; COAPEHUM, 2012); five LIS programs are among them (Table 13.5).

In summary, from a total of eight undergraduate LIS programs, six have earned a Class 1 status and five have been accredited (Table 13.6).

13.2.4   Evaluation with Accreditation Purposes: Graduate Programs

Another important initiative to quality improvement is the National Program for Quality Graduate Studies (PNPC), sponsored by the National Council for Science and Technology (CONACYT) and SEP. This initiative from the public sector acknowledges graduate programs with "highly regarded academic staff, high rates

Table 13.7 PNPC's criteria for accreditation (Mexico)

Category	Criterion	Number of sub-criteria
Program structure	Curriculum	10
	Teaching–learning process	2
Students	Admission	1
	Track records	
	Mobility	
	Tutorship and counseling	
	Full-time students	
Faculty	Staff affairs	6
	Knowledge creation/application	2
Infrastructure	Venues and equipment	2
	Laboratories and workshops	3
	Information and library resources	2
	Information and communication technologies	3
Program outcomes	Program scope and evolution	2
	Program relevance	2
	Program effectiveness	1
	Contribution to knowledge	8
External cooperation	Academic cooperation	2
	Funding	2
Total	19	48

of graduation, adequate infrastructure and high scientific or technological productivity." Graduate programs with these characteristics are listed on the PNPC database (CONACYT, 2013).

Graduate programs willing to enter the PNPC list are evaluated according to the PNPC framework, which takes into account 19 criteria, organized in six areas (CONACYT, & SEP, 2011, 21–23) (Table 13.7).

After examining statistical data, self-evaluation reports, and peer evaluation reports, the National Council of Graduate Studies (CNP) ponders whether to admit or not a graduate program to the PNPC list. If admitted, the graduate program is placed in one of the following categories:

- Category IV (internationally recognized): programs with established agreements with foreign peer institutions in terms of faculty and student mobility, cosupervision of theses and dissertations, and joint research projects.
- Category III (developed): nationally recognized programs that are socially relevant, academically productive, effective in educating qualified professionals, and closely related with other actors in society.
- Category II (developing): programs with promising academic prospects with realistic goals that can be achieved in the medium term.
- Category I (recently created): doctoral programs with no more than 4.5 years of operation or master's programs with no more than 2.5 years, that comply with the basic criteria of the PNPC framework.

As can be seen in Appendix 1, there are four graduate LIS programs in Mexico. The ones offered by the National Autonomous University of Mexico (UNAM) have been evaluated by the PNPC. The master's degree in Library and Information Studies and the PhD in Library and Information Studies were put in Category III (developed) and Category IV (internationally recognized) respectively.

13.3   Colombia

The Colombian quality assurance system in higher education is implemented at two levels. At the bottom level, the National Commission of Quality Assurance in Higher Education (CONACES) verifies that new academic programs comply with minimum requirements in order to be officially recognized. At the top level, the National Council of Accreditation (CNA) evaluates HEIs and programs, and recommends that those that meet the highest quality standards in education be accredited (CNA, 2009, 9–10). Whereas at the bottom level reviews are mandatory (i.e., programs that do not meet the criteria established by CONACES cannot be delivered), at the top level evaluations are voluntary (CNA, 2012a, 8; Ministerio de Educación Nacional, 2010). In both cases, the pronouncements about the results are made by the Ministry of National Education, taking into account the final evaluation reports presented by CONACES and CNA.

CNA is part of the National System of Accreditation (SNA, for its initials in Spanish), which comprises "policies, strategies, processes, and organizations that seek to assure the society that higher education institutions meet the highest quality standards and accomplish their goals and objectives" (CNA, 2012b). The SNA was officially established in 1992 by the law that regulates higher education in Colombia (*Ley 30*). This law also established the National Council of Higher Education (CESU) with the mission to implement the SNA and organize CNA. As such, CESU is responsible for the design of policies on accreditation.

The SNA constitutes a joint effort between the public and the private sector. On the side of the government, accreditation procedures are regulated by the Colombian law and funded by the State, accreditation policies are given by CESU, and results are promulgated by the Ministry of Education. On the other hand, HEIs are responsible for conducting self-evaluations and preparing improvement plans, while academic peers are accountable for performing external reviews and making judgments (Roa Varelo, 2005, 153).

The HEIs being evaluated can track the status of the process through an information system called Quality Assurance System in Higher Education (SACES). This online tool can also be consulted by other actors, such as government officials, academic peers participating in a review process, researchers needing to examine statistical data and indicators in higher education, or students choosing a school or academic program.

13.3.1   Minimum Requirements

Established in 2003, CONACES is the official agency that, among other responsibilities commissioned by the Ministry of National Education, certifies that new programs meet certain requirements at an adequate level. This evaluation also applies to programs that have reformulated their academic offering. As was mentioned before, this is a mandatory procedure; all academic programs must be examined by CONACES and in fact, this represents the first step towards accreditation. Programs that obtain approval enter a database of officially qualified higher education offerings called National Information System on Higher Education (SNIES) and are given an identification code.

New or modified programs are evaluated against aspects that include curriculum, students (admission and assessment), faculty, alumni, infrastructure, administration, and financial resources. CONACES also requires that HEIs meet some institutional conditions. The evaluation is conducted by a group of peers selected by CESU after a public call. If the evaluation results are favorable, CONACES issues a certification that is valid for 7 years (Ministerio de Educación Nacional, 2013).

13.3.2   Accreditation Model

Evaluating quality in education involves identifying characteristics that distinguish a program or an institution and judging the relative distance between those current characteristics and the ones that the optimum would have. To assess how far or close that optimum is, CNA has defined a set of general characteristics which serve as a reference, where the specific relative weights depend on the nature of the HEI and the program under evaluation (CNA, 2012a, 14). Both types of accreditation (institutional and programmatic) are conducted by CNA.

For this purpose, CNA has made explicit its ethical position on the accreditation process. Its model is based on 11 principles (universality, integrity, equity, suitability, responsibility, coherence, transparency, pertinence, efficacy, efficiency, and sustainability) that as a whole, provide a value-based framework for judgments about quality (CNA, 2012a, 9–11). The academic peers involved in the process are also expected to abide by the CNA code of ethics (CNA, 2012c).

13.3.3   Programmatic Accreditation

The accreditation of academic programs goes through five stages:

1. Initial conditions. CNA verifies that certain formal requirements are met.
2. Self-evaluation. The HEI evaluates the academic program according to the criteria of the CNA model.

3. External evaluation. A group of designated peers verify that the self-evaluation process has been rigorously followed, visit the HEI facilities, meet with its authorities, and judge the quality of program. If necessary, peers also make recommendations and suggestions for improvement.
4. Final evaluation. Based on the self-evaluation results and the external peer review, CNA issues a final report.
5. Public acknowledgment. Based on the CNA final report, the Ministry of National Education officially recognizes that the program meets high quality standards and confers the status of accredited on the academic program.

Since these official recognitions are not permanent (valid 4 years at a minimum and 9 years at a maximum), accredited programs should submit themselves to new accreditation processes to renew their conditions as such.

According to the CNA model's nomenclature and hierarchy, the self-evaluation and the external peer review are performed taking into account ten areas de evaluation (factors). Each factor is examined against a number of characteristics (40 in total), which in turn can be assessed in terms of aspects (251 in total). The following table shows the characteristics grouped in factors and the number of aspects associated to each characteristic (CNA, 2012a, 17–53) (Table 13.8):

Table 13.8 CNA's criteria for programmatic accreditation (Colombia)

Factor	Characteristic	Number of aspects
Institutional and program projects	Institutional mission and strategic direction	6
	Program educational project	4
	Program academic and social relevance	9
Students	Selection and admission	4
	Admitted students and institutional capacity	7
	Extracurricular activities	3
	Student and academic regulations	5
Faculty	Selection, hiring, and retention	3
	Teaching career regulations	6
	Number of full-time and part-time faculty, academic qualifications, and relevant experience	8
	Pedagogical training and professional development courses	6
	Institutional encouragement to good practices in teaching, research, artistic/cultural creation, reach out initiatives, and international cooperation	3
	Creation, application and impact of teaching materials	4
	Merit-based salaries	3
	Evaluation	5

(continued)

Table 13.8 (continued)

Factor	Characteristic	Number of aspects
Academic processes	Curriculum integrality	10
	Curriculum flexibility	10
	Interdisciplinary approach	3
	Teaching–learning methodologies	14
	Student assessment system	6
	Student portfolio	5
	Program evaluation and self-regulation	4
	Reach out initiatives	8
	Library resources	5
	Information and communication technologies	6
	Facilities for teachers	6
National and international visibility	Program recognition in national and international academic contexts	9
	Faculty and student mobility	8
Research, innovation, and artistic/ cultural creation	Education intended for research, innovation, and artistic/ cultural creation	11
	Program commitment to research, innovation, and artistic/ cultural creation	8
Institutional well-being	Institutional well-being policies, initiatives, and services	9
	Student retention	3
Organization and administration	Program organization and management	7
	Information and communication systems	9
	Staffing and leadership	4
Alumni	Tracking	8
	Contribution to society and the academic world	4
Physical and financial resources	Facilities	5
	Funding	8
	Resource administration	5
Total	40	251

13.3.4   Institutional Accreditation

This procedure examines an institution as a whole against ten factors: institutional mission and strategic direction, academic community (students, faculty, and researchers), academic processes (teaching, research, and extension activities), research, institutional well-being, social relevance and impact on society, self-evaluation and self-regulation, governance and management, infrastructure and academic support resources, and financial resources (CNA, 2006, 22).

The CNA understands that, in providing education services, HEIs also fulfill distinct conceptions about how education should be delivered, according to their own systems of ideals, values, and expectations. That is why, in the case of institutional

Table 13.9 Accredited LIS programs (Colombia)

HEI	Year of award: institutional accreditation	Academic program	Year of award: programmatic accreditation
Pontifical Javeriana University (PUJ)	2003, 2012	Information Science—Librarianship	2002, 2006
University of Antioquia (UDEA)	2003, 2012	Library Science	1999, 2004
University of La Salle (ULSalle)	2008, 2012	Information and Documentation Systems	2007, 2012

accreditation, the CNA model evaluates quality not only in terms of universally desirable characteristics, but also considering those ones that make one HEI particularly different from the others. Those traits are assessed by examining mission statements and strategic plans (CNA, 2006, 13).

The last official report shows that, as of May 2011 and since 2003, the CNA has accredited 26 HEIs in Colombia (CNA, 2011).

13.3.5   Accredited LIS Programs and HEIs

In Colombia, LIS programs are offered by four HEIs (see Appendix 2). Three of them, along with the LIS programs they deliver, have gotten accredited (Table 13.9).

The experience of the LIS school at the University of Antioquia has been documented by Jaramillo (2005) and Múnera Torres (2010). Jaramillo (2005, 16) reports that the Inter-American School of Library Science (EIB) at the University of Antioquia (UDEA) formally started a self-evaluation process in 1996. This self-evaluation "not only [seeked] to make explicit [the school's] goals of [high degree of] development, and more efficacy and efficiency for improvement purposes, but also to support [its] commitment to educating highly qualified professionals [...] and to becoming more competitive at the national and at the international level." In 1999, after its Library Science program earned its first accreditation, the EIB executed a quality assurance plan along three lines of action: the consolidation of its strengths, the overcoming of its weaknesses, and the demonstration of its ability to innovate. In 2004, the Ministry of National Education renewed its accreditation for 7 years. Múnera Torres (2010, 197) makes an account of the benefits that the accreditation brought to the EIB.

Múnera Torres and Giraldo Giraldo (2011, 189) report that the Information Science—Librarianship program at the Pontifical Javeriana University (PUJ) got first accredited in 2002 (for a four-year period), an status that was later confirmed in 2006. The authors highlight that, over the years, all the actors involved in the academic program have developed a culture of constant self-regulation and evaluation, a culture which is supported by the institutional apparatus. From 2001 to 2004, a designated committee conducted meetings, surveys, and self-evaluation exercises among groups of faculty, students, and alumni. It is worth mentioning that this committee was integrated not only by the faculty directly involved, but also by one student and the authorities of the upper academic unit (i.e., the Faculty of Communication and Languages). The results obtained from this initiative served as inputs for making adjustments in the program in terms of contents, pedagogical techniques, delivery methods, etc. as a measure to counteract its weaknesses. Other institutional quality assurance mechanisms are systematic reviews performed by the university's Vice-Rectorate of Academic Affairs and the special monitoring group of the Faculty of Communication and Languages.

The same authors also recount that the Information and Documentation Systems program at the University of La Salle (ULSalle) undertook an evaluation in 2005, exercise that would successfully conclude with the accreditation conferred in 2007. This status was recently confirmed in 2012 for a subsequent 6-year period. As in the case of PUJ, the self-evaluation was fully supported by the top-level authorities of ULSalle. In line with the ULSalle's Institutional Plan of Development (2010–2015), which considers the Continuous Quality Improvement Plan as one of its strategic programs, the self-evaluation at ULSalle led to three concrete measures. First, after a thorough review, a new student-centered, interdisciplinary, technology-intensive, and socially oriented curriculum was devised. Second, a more numerous and consolidated full-time faculty has allowed the department to increase its academic production in terms of quantity and quality. The sponsorship programs that encourage faculty to take more advanced academic degrees have been instrumental in this regard. Finally, the accreditation process also triggered the reconceptualization of the program in relation to research, steering away from a teaching-centered program towards one that learns through research (Múnera Torres & Giraldo Giraldo, 2011, 198–200).

Regarding institutional accreditations, the three Colombian universities that offer face-to-face LIS programs have won this distinction. Following the recommendations made by CNA, the Ministry of National Education renewed their status of high quality HEIs last year. In March 2012, PUJ was reaccredited for 8 years (its first accreditation being in 2003 for 9 years); in December, both UDEA and ULSalle were given the same award, for 9 and 6 years, respectively. UDEA had previously earned the accreditation in 2003 for a 9-year period, and ULSalle in 2008 for a 4-year period.

13.4 Costa Rica

In Costa Rica, accreditations are officially conferred to undergraduate and graduate programs by the National System of Accreditation in Higher Education (SINAES, for its acronym in Spanish), a nongovernmental autonomous organization, acknowledged in 2002 by the 8256 Law (Mora Alfaro, 2004, 109). SINAES brings together public as well as private universities that voluntarily have decided to adhere to its "principles of quality and academic excellence" and internal regulations (SINAES, 2009, 8). In fact, new members are expected to obtain accreditation for at least one program within the 2 years following their affiliation.

SINAES currently has 19 members (HEIs) that, as a whole, have a student population of 100,000, enrolled in undergraduate and graduate programs (SINAES, 2013).

As the official agency and the highest authority regarding the accreditation of undergraduate and graduate programs of higher education, SINAES seeks to "plan, organize, develop, implement, control, and monitor official accreditation processes that continuously guarantee the quality of programs [...] offered by Higher Education Institutions, and treat records with complete confidentiality" (SINAES, 2009, 9). SINAES aims at conducting its operations based on excellence, integrity, social responsibility, respect, and leadership (SINAES, 2009, 8–9).

SINAES has also been recognized by the Central American Council of Accreditation in Higher Education (CCA) which enables it to sign cooperation agreements with similar agencies of other countries in the region. In 2010 SINAES became the first Latin American organization recognized by the International Network for Quality Assurance Agencies in Higher Education (INQAAHE), as an institution that "comprehensively adheres to the INQAAHE Guidelines of Good Practice (GGP)" (INQAAHE, 2013).

As mentioned, SINAES was acknowledged and authorized by the Costa Rican government in 2002. However, Mora Alfaro (2004, 108) indicates that SINAES started its operations in 1999, as a joint effort of four public universities (The University of Costa Rica, the Costa Rica Institute of Technology, the National University of Costa Rica, and the State University of Distance Education.). These universities, concerned about the development of higher education, joined together to respond to "the increasing social demand for an efficient quality assurance mechanism" in "a world immersed in an accelerated process of economic, political, and cultural integration." In 1974 the same four public universities had signed an agreement to create the National Council of Rectors (CONARE), a body comprised of their highest authorities (CONARE, 2013).

In terms of organization, SINAES has a National Council of Accreditation, as the central executive organ that defines the strategic direction the agency will take. This council is comprised of eight highly regarded academics and professionals who do not hold any authority position neither in the Costa Rican government nor at any HEI (Alvarado, 2005, 144).

It should be mentioned that the Costa Rican institutional platform is formed by 80 HEIs, 56 of which are universities (five public and 51 private) (Lemaitre & Zenteno, 2012, 60). There is not an official agency that records statistics about programs in higher education. However, it is estimated that 1,139 "academic opportunities" were offered in 2010 (Brunner & Ferrada Hurtado, 2011, 136–137); 69 from this group have been accredited by SINAES, as of December 2012 (SINAES, 2012).

Besides the procedures established by SINAES, there are alternative quality assurance mechanisms in the country. SINAES coexists with the Accreditation System of the Private Higher Education Teaching in Costa Rica (SUPRICORI), a nonofficial accreditation agency born as an initiative of private universities (Lemaitre & Zenteno, 2012, 36). It was founded in 2001 by the Unit of Private Universities Rectors of Costa Rica (UNIRE) (Barboza Jiménez & Rodríguez Salas, 2011, 126). Currently, 32 private universities are affiliated to UNIRE. Other accreditation practices in the country include the submission of academic programs to external revision by organizations in the Central American region (SICAR/CSUCA: Central American Superior University Council) or beyond (CEAB: Canadian Engineering Accreditation Board, SACS: Southern Association of Colleges and Schools and EQUIS: European Quality Improvement System). These practices of quality assurance were common even before SINAES was established (Lemaitre & Zenteno, 2012, 60). Some universities, especially those that are public, have their own internal quality assurance procedures. This is the case of the Center of Academic Evaluation at the University of Costa Rica.

Finally, it should be noted that all accreditation processes in Costa Rica are focused on academic programs; there are no mechanisms for institutional accreditation (Lemaitre & Zenteno, 2012, 60).

13.4.1 Accreditation Model

The model followed by SINAES seeks to "visualize, in an integrating way, the main elements involved in the educational process: the background that serves as a context, the resources and consumables necessary for the process, the process itself and the results." According to the SINAES nomenclature, these four dimensions (i.e., context, resources, process, and results) are examined in terms of components that are particular to the program under evaluation (SINAES, 2009, 38).

The SINAES model also takes into consideration four complementary components. The first two are part of the evaluation itself: admissibility (i.e., which requirements in the Costa Rican higher education system must mandatorily be met before the accreditation procedure starts) and sustainability (i.e., how programs will be able to maintain their accredited status). The other two elements, as external components, include guidelines for academic units to conduct meta-evaluations (i.e., examine how the evaluation was carried out in terms of participation, institutional support, planning, sustainability, and information management) and to prepare for future reaccreditations (i.e., monitor how closely the program has followed the improvement plan previously set) (SINAES, 2009, 39–40).

Table 13.10 shows the 21 units of evaluation (components) of the SINAES model, arranged in four dimensions. It also shows the number of criteria (particular aspects) and—if possible—standards (acceptable levels of quality), which serve as comparison indicators for evaluating each component; and the number of evidence pieces, which make the components observable (SINAES, 2009, 42).

The accredited status is valid for a period of 4 years at most. The task of performing the external evaluation is entrusted to a panel of academic peers—Costa Rican and foreign nationals—with no connection with the HEI whose program is under evaluation. Peers from Chile, Colombia, Mexico, the USA, Peru, Uruguay, Puerto Rico, Argentina, and Spain have been invited by SINAES (Mora Alfaro, 2004, 110).

Table 13.10 SINAES' criteria for accreditation (Costa Rica)

Dimension	Component	Criterion	Standard	Evidence
	Admissibility	12	0	19
Relation to the context	Information and promotion	2	2	4
	Admission procedure	2	0	5
	Correspondence to the context	6	1	11
Resources	Curriculum	19	1	38
	Faculty	12	7	27
	Administrative staff	4	0	9
	Infrastructure	8	1	17
	Information and resource center	5	2	18
	Equipment and materials	5	2	10
	Finance and budgeting	2	0	4
Process	Faculty development	5	1	16
	Teaching and learning methodologies	7	1	8
	Management	20	3	39
	Research	9	1	21
	Extension activities	9	1	15
	Student life	17	3	38
Results	Student performance	5	0	11
	Alumni	11	5	28
	Profession's projection	1	3	6
	Sustainability	10	0	4
Total	21	171	34	348

13.4.2   Accreditation Process

The accreditation process for academic programs, according to the model followed by SINAES, goes through four stages:

1. Initial stage. The program academic community formally takes the decision to submit the program to evaluation, decision that is supported by the HEI's authorities.
2. Self-evaluation. The actors involved (faculty, administrative staff, students, alumni, employers, professional associations, etc.) evaluate and reflect upon the program according to established criteria.
3. External evaluation. Designated peers validate the self-evaluation process and the self-evaluation report, and perform an evaluation in situ.
4. Accreditation and continuous improvement. The National Council of Accreditation, after examining the final external evaluation report prepared by the peers, and the self-evaluation report and the *Commitment to Improvement* document prepared by the HEI, decides whether to confer the accreditation or not. After granting the accreditation, SINAES periodically monitors how the goals established on the *Commitment to Improvement* document are being progressively met.

Table 13.11 Accredited LIS programs (Costa Rica)

HEI	Academic program	Year of award
National University of Costa (UNACR)	Librarianship and documentation	2005, 2010[a]

A complete list of HEIs and programs is shown in Appendix 3

[a]This year is an estimation. The program has been reaccredited as shown on the SINAES Web site. However, the exact year of the reaccreditation could not be found

13.4.3   Accredited LIS Programs

The only LIS program accredited by SINAES is offered by the National University of Costa Rica (SINAES, 2012) (Table 13.11).

Efforts towards the accreditation at the National University of Costa Rica first started in 2003. It was a decision formally taken by the School of Librarianship, Documentation and Information, as stated by Miranda Arguedas (2005, 118). "The self-evaluation made it possible to revise the everyday operations at the school, to question paradigms and to make adjustments in the program to adapt it to social demands. In order to achieve that it was necessary to gather opinions from all the actors involved—internal and external to the school. For this purpose, students, faculty, alumni, and employers took part in a series of meetings and workshops where strengths and weaknesses concerning the school's curriculum, logistics support, and human resources could be identified." On accounting the experience, Miranda Arguedas (2005, 118) also indicates how the school benefited from the self-evaluation even before the peer review, most notably through the redesign of the curriculum, the standardization of working conditions and assessment procedures for the faculty, and the improvement in infrastructure and equipment. The designated academic peers visited the school in April 2005 (Miranda Arguedas, 2005, 120) and after a positive review, the SINAES conferred the accreditation that year in August (Rodríguez Salas, 2010, 185).

The quality cycle started again in 2008, with a new self-evaluation process towards reaccreditation. As pointed out by Rodríguez Salas (2010, 178), on that second occasion, more actors were involved. Apart from students, faculty, employers, and alumni, the revision also included surveys and interviews with the school's authorities and administrative staff. Among the favorable impacts that stemmed from submitting the school to internal and external evaluation, Rodríguez Salas (2010, 181–184) mentions improvements in management, infrastructure, research, and curriculum. She finally reflects on how the insights gained during the process served as a catalyst to make the academic and administrative staff aware of their role in building a successful academic offer and in the end, to promote an organizational culture based on quality.

13.5   Final Thoughts

This article has presented the mechanics of accreditation in three countries of the Latin American region considered to exhibit well-developed national quality assurance systems. Although similar to each other in some aspects (e.g., non-mandatory nature, rather comparable set of accreditation criteria), these systems differ in other angles. Most notably, accreditation systems originated from HEIs in Mexico and Costa Rica, and the government in Colombia; are decentralized in Mexico, relatively more centralized in Costa Rica, and centralized in Colombia; and were found to be more tightly regulated by the government in Colombia. (A list of quality assurance organizations is shown in Appendix 4.)

In a region where education is considered a public good and HEIs are expected to be accountable for the quality of education they provide, it was reassuring to find well-established accreditation practices in LIS education. Unfortunately, this does not seem to be the norm. Until more LIS schools embrace formal quality assurance procedures, this will remain pending in their agendas.

Appendix 1: LIS Programs in Mexico

Associate's degree

	HEI	Academic program
1	National School of Library and Archival Science (ENBA)	Library Science (5 semesters)

Bachelor's degree

	HEI	School	Academic program
1	Autonomous University of Chiapas (UNACH)	Faculty of Humanities	Library Science and Information Management (9 semesters)
2	Autonomous University of Chihuahua (UACH)	Faculty of Philosophy and Letters	Information Science (9 semesters)
3	Autonomous University of Mexico State (UAEMex)	Faculty of Humanities	Documentary Information Sciences (10 semesters)
4	Autonomous University of Nuevo Leon (UANL)	Faculty of Philosophy and Letters	Library and Information Science (10 semesters)
5	Autonomous University of San Luis Potosi (UASLP)	School of Information Science	Library and Information Studies (8 semesters)
6	National Autonomous University of Mexico (UNAM)	Faculty of Philosophy and Letters, College of Librarianship	Library and Information Studies (8 semesters)
7	National School of Library and Archival Science (ENBA)		Library Science (9 semesters)

Master's degree

HEI	School	Academic program
1 Autonomous University of Chihuahua (UACH)	Faculty of Philosophy and Letters	Library and Information Science (4 semesters, full-time)
2 National Autonomous University of Mexico (UNAM)	Faculty of Philosophy and Letters, College of Librarianship	Library and Information Studies (4 semesters, full-time)
3 The College of Mexico		Library Science (4 semesters, full-time)

Doctoral degree

HEI	School	Academic program
1 National Autonomous University of Mexico (UNAM)	Faculty of Philosophy and Letters, College of Librarianship	Library and Information Studies (8 semesters, full-time)

Appendix 2: LIS Programs in Colombia

Bachelor's degree

HEI	School	Academic program
1 Pontifical Javeriana University (PUJ)	Faculty of Communication and Languages	Information Science—Librarianship (10 semesters)
2 University of Antioquia (UDEA)	Inter-American School of Library Science	Library Science (8 semesters)
3 University of La Salle (ULSalle)	Faculty of Economics and Social Sciences	Information and Documentation Systems (10 semesters)
4 University of Quindio	Faculty of Human Sciences and Fine Arts	Information Science, Documentation, Library and Archive Science (10 semesters, online)

Appendix 3: LIS Programs in Costa Rica

Intermediate Bachelor's degree

HEI	School	Academic program
1 Autonomous University of Central America (UACA)	Faculty of Human Sciences, School of Librarianship	Librarianship (4 years)
2 National University of Costa Rica (UNACR)	Faculty of Philosophy and Letters, School of Librarianship, Documentation and Information	Librarianship and Documentation (4 years), with emphasis on: • Information management, or • Information and communication technologies
3 University of Costa Rica (UCR)	Faculty of Education, School of Library and Information Science	Librarianship (4 years), with emphasis on: • Information Science, or • Educational Libraries

Bachelor's degree

HEI	School	Academic program
1 National University of Costa Rica (UNACR)	Faculty of Philosophy and Letters, School of Librarianship, Documentation and Information	Librarianship and Documentation (6 years), with emphasis on: • Information management, or • Information and communication technologies
2 University of Costa Rica (UCR)	Faculty of Education, School of Library and Information Science	Library and Information Science (6 years)
3 State University of Distance Education (UNED)	School of Social Sciences and Humanities	Librarianship, educational libraries and learning resources centers (6 years) Librarianship and information and communication technologies (6 years)

Master's degree

HEI	School	Academic program
1 University of Costa Rica (UCR)	Faculty of Education, School of Library and Information Science	Library and Information Studies

Appendix 4: Quality Assurance Organizations

Quality assurance organizations and programs in Mexico

ACCECISO	Association for the Accreditation and Certification in Social Sciences
ANUIES	National Association of Universities and Higher Education Institutions
CENEVAL	National Center for Evaluation in Higher Education
CIEES	Inter-institutional Committees for Evaluation in Higher Education
COAPEHUM	Council for the Accreditation of Academic Programs in Humanities
CONACYT	National Council for Science and Technology
COPAES	Council for the Accreditation of Higher Education
PNPC	National Program for Quality Graduate Studies
SEP	Ministry of Public Education
SNI	National System of Researchers

Quality assurance organizations and information systems in Colombia

CESU	National Council of Higher Education
CNA	National Council of Accreditation
CONACES	National Commission of Quality Assurance in Higher Education
SACES	Quality Assurance System in Higher Education
SNA	National System of Accreditation
SNIES	National Information System on Higher Education

Quality assurance organizations in Costa Rica

CONARE	National Council of Rectors
SINAES	National System of Accreditation in Higher Education
SUPRICORI	Accreditation System of the Private Higher Education Teaching in Costa Rica
UNIRE	Unit of Private Universities Rectors of Costa Rica

References

ACCECISO. (2012). Programas acreditados por ACCECISO del 2003 a la fecha. Retrieved February 26, from http://www.acceciso.org.mx/esp/acr.html.

Alvarado, M. (2005). La experiencia de Costa Rica. In CINDA (Ed.), *Los procesos de acreditación en el desarrollo de las Universidades* (pp. 143–150). Santiago de Chile: IESALC/UNESCO.

ANUIES. (1997). La evaluación y acreditación de la educación superior en Mexico. *Revista de la Educacion Superior 26*(101). Retrieved, from http://publicaciones.anuies.mx/revista/101/3/2/es/evaluacion-y-acreditacion-de-la-educacion-superior-en-mexico.

ANUIES. (2006). Estatuto de la Asociación Nacional de Universidades e Instituciones de Educación Superior de la República Mexicana, A.C. México DF.

ANUIES. (2012). Acerca de la ANUIES. Retrieved December 19, from http://www.anuies.mx/content.php?varSectionID=2.

Barboza Jiménez, L., & Rodríguez Salas, K. (2011). Entes Acreditadores Centroamericanos, Regionales e Iberoamericanos. In L. Escalona Ríos (Ed.), *La Evaluación de la Educación Bibliotecológica en América Latina* (pp. 85–138). México, DF: UNAM, Centro Universitario de Investigaciones Bibliotecológicas. Retrieved, from http://132.248.242.3/~publica/archivos/libros/evaluacion_educacion_bibliotecologica.pdf.

Brunner, J. J., & Ferrada Hurtado, R. (Eds.). (2011). *Educación Superior en Iberoamérica: Informe 2011*. Santiago de Chile: RIL editores, Universia, CINDA.

CIEES. (2001). Criterios Para Evaluar Programas Académicos de Licenciatura y Posgrado. México, DF. Retrieved February 26, from http://csh.izt.uam.mx/licenciaturas/psicologia_social/comision/criterios_ciees.pdf.

CIEES. (2013). Consultas de Programas Evaluados por los CIEES al 31 de Enero de 2013. Retrieved February 7, from http://www.ciees.edu.mx/ciees/reportesCmysql/consultas.htm.

CNA. (2006). *Lineamientos para la Acreditación Institucional*. Bogotá: Ministerio de Educación Nacional. Retrieved February 26, from http://cms-static.colombiaaprende.edu.co/cache/binaries/articles-186359_lineamientos_3.pdf?binary_rand=2342.

CNA. (2009). *Lineamientos para la Acreditación de Alta Calidad de Programas de Maestría y Doctorado*. Bogotá: Ministerio de Educación Nacional. Retrieved February 26, from http://cms-static.colombiaaprende.edu.co/cache/binaries/articles-186363_archivo_pdf_lineamientos_MyD_final.pdf?binary_rand=2245.

CNA. (2011). *Evolución de los Procesos de Acreditación Institucional 2003-2011*. Retrieved February 26, from http://www.cna.gov.co/1741/articles-188924_recurso_3.pdf.

CNA. (2012a). *Lineamientos para la Acreditación de Programas de Pregrado*. Bogotá: Ministerio de Educación Nacional. Retrieved, from http://www.cna.gov.co/1741/articles-186359_pregrado_2012.pdf.

CNA. (2012b). *Sistema Nacional de Acreditación en Colombia*. Retrieved February 26, from http://www.cna.gov.co/1741/article-186365.html.

CNA. (2012c). *Pares Académicos*. Retrieved February 26, from http://www.cna.gov.co/1741/article-186793.html.

COAPEHUM. (2012). *Actividades COAPEHUM 2008-2012*. Retrieved February 26, from http://www.coapehum.org/reportes.html.

CONACYT. (2013). *Programa Nacional de Posgrados de Calidad*. Retrieved February 26, from http://www.conacyt.gob.mx/FormacionCapitalHumano/Paginas/PosgradosCalidad.aspx.

CONACYT, & SEP. (2011). *Programa Nacional de Posgrados de Calidad. Marco de Referencia para la Evaluación y Seguimiento de Programas de Posgrado*. México, DF: CONACYT. Retrieved, from http://www.conacyt.gob.mx/FormacionCapitalHumano/Documents/PNPC/Marco_Referencia_Escolarizada.pdf.

CONARE. (2013). *Consejo Nacional De Rectores—CONARE*. Retrieved February 26, from http://www.conare.ac.cr/.

COPAES. (2012). *Marco General para los Procesos de Acreditaciónd de Programas Académicos del Nivel Superior*. México, DF. Retrieved, from http://www.copaes.org.mx/FINAL/docs/MARCO_DE_REFERENCIA_COPAES_2012.pdf.

COPAES. (2013). *Organismos Acreditadores*. Retrieved February 26, from http://www.copaes.org.mx/FINAL/organismos_acred.php.

Escalona Ríos, L. (Ed.). (2011). *La Evaluación de la Educación Bibliotecológica en América Latina*. México, DF: UNAM, Centro Universitario de Investigaciones Bibliotecológicas. Retrieved, from http://132.248.242.3/~publica/archivos/libros/evaluacion_educacion_bibliotecologica.pdf.

Fresán Orozco, M., & Taborga Torrico, H. (Eds.). (2002). *Indicadores y Parámetros para el Ingreso y la Permanencia de Instituciones de Educación Superior a la ANUIES*. México, DF: ANUIES.

INQAAHE. (2013). *GGP Aligned Agencies*. Retrieved February 26, from http://www.inqaahe.org/main/professional-development/guidelines-of-good-practice-51/ggp-aligned-agencies.

Jaramillo, O. (2005). El Proceso de Acreditación de los Programas de Formación Profesional en Bibliotecología. *Ciencias de la Información, 36*(1), 13–37.

Lemaitre, M. J., & Zenteno, M. E. (Eds.). (2012). *Aseguramiento de La Calidad en Iberoamérica. Educación Superior: Informe 2012*. Santiago de Chile: Centro Interuniversitario de Desarrollo, CINDA; Universia. Retrieved, from http://www.cinda.cl/download/CINDA_2012_Informe_de_Educacion_Superior.pdf.

Martínez Rider, R. M. (2010). Los Procesos de Evaluación y de Acreditación en la Escuela de Ciencias de la Información de la Universidad Autónoma de San Luis Potosí. In J. Ríos Ortega & J. J. Calva González (Eds.), *Memoria del XXVII Coloquio de Investigación Bibliotecológica y sobre la Información, 28-30 de Septiembre de 2009: La Investigación y la Educación Bibliotecológica en la Sociedad del Conocimiento* (pp. 241–250). México, DF: UNAM, Centro Universitario de Investigaciones Bibliotecológicas. Retrieved, from http://132.248.242.3/~publica/archivos/libros/xxvii_coloquio_cuib.pdf.

Ministerio de Educación Nacional. (2010). Sistema de Aseguramiento de la Calidad de la Educación Superior. Retrieved August 20, from http://www.mineducacion.gov.co/1621/w3-article-235585.html.

Ministerio de Educación Nacional. (2013). ¿Qué Es SACES?. Retrieved February 26, from http://www.mineducacion.gov.co/sistemasdeinformacion/1735/article-221614.html.

Miranda Arguedas, A. (2005). Situación y Perspectiva de la Educación Bibliotecológica en Costa Rica. In F. F. Martínez Arellano & J. J. Calva González (Eds.), *Seminario INFOBILA como Apoyo a la Investigación y Bibliotecologia en América Latina y El Caribe* (pp. 108–132). México, DF: UNAM, Centro Universitario de Investigaciones Bibliotecológicas.

Mora Alfaro, J. (2004). El Sistema de Acreditación de la Educación Superior de Costa Rica. In *La Evaluación y la Acreditación en la Educación Superior en América Latina y El Caribe* (pp. 99–113). Caracas: IESALC/UNESCO (UNESCO's International Institute for Higher Education in Latin America and the Caribbean). Retrieved, from http://www.iesalc.unesco.org.ve/dmdocuments/biblioteca/libros/EvalyAcredALC.pdf.

Múnera Torres, M. T. (2010). Impacto de los Procesos de Acreditación en la Escuela Interamericana de Bibliotecología de la Universidad de Antioquia. In J. Ríos Ortega & J. J. Calva González (Eds.), *Memoria del XXVII Coloquio de Investigación Bibliotecológica y sobre la Información, 28-30 de Septiembre de 2009: La Investigación y la Educación Bibliotecológica en la Sociedad del Conocimiento* (195–203). México, DF: UNAM, Centro Universitario de Investigaciones Bibliotecológicas. Retrieved, from http://132.248.242.3/~publica/archivos/libros/xxvii_coloquio_cuib.pdf

Múnera Torres, M. T. (2012). Procesos de Aseguramiento de la Calidad en la Formación Bibliotecológica de América del Sur. *Revista Interamericana de Bibliotecología, 35*(1), 63–72.

Múnera Torres, M. T., & Giraldo Giraldo, Y. N. (2011). Los Procesos de Evaluación Educativa en América del Sur. In L. Escalona Ríos (Ed.), *La Evaluación de la Educación Bibliotecológica en América Latina* (pp. 189–201). México, DF: UNAM, Centro Universitario de Investigaciones Bibliotecológicas. http://132.248.242.3/~publica/archivos/libros/evaluacion_educacion_bibliotecologica.pdf.

Rama, C. (2011). La Educación Superior en América Latina en el Periodo 2000–2010: Ocho Ejes Centrales en Discusión. *Innovación Educativa 57*(11), 15–20. Retrieved, from http://www.redalyc.org/redalyc/src/inicio/ArtPdfRed.jsp?iCve=179422350003.

Roa Varelo, A. (2005). La Experiencia de Colombia. In CINDA (Ed.), *Los Procesos de Acreditación en el Desarrollo de las Universidades* (pp. 151–161). Santiago de Chile: IESALC/UNESCO.

Rodríguez Salas, K. (2010). Una Mirada Crítica al Proceso de Autoevaluación y Acreditación en la Escuela de Bibliotecología, Documentación e Información. In J. Ríos Ortega & J. J. Calva González (Eds.), *Memoria del XXVII Coloquio de Investigación Bibliotecológica y sobre la Información 28-30 de Septiembre de 2009. La Investigación y la Educación Bibliotecológica en la Sociedad del Conocimiento* (pp. 177–186). México, DF: UNAM, Centro Universitario de Investigaciones Bibliotecológicas. Retrieved, from http://132.248.242.3/~publica/archivos/libros/xxvii_coloquio_cuib.pdf.

Rubio Oca, J. (2007). La Evaluación y Acreditación de la Educación Superior en México: Un Largo Camino Aún por Recorrer. *Reencuentro* diciembre (50), 35–44.

SINAES. (2009). Manual de Acreditación Oficial de Carreras de Grado del Sistema Nacional de Acreditación de La Educación Superior. San José. Retrieved, from http://www.sinaes.ac.cr/manual_guias/manual_oficial_acreditacion_vf_feb2010.pdf.

SINAES. (2012). Carreras con Garantía de Calidad. San José. Retrieved February 26, from http://www.sinaes.ac.cr/carreras_acreditadas/carreras.pdf.

SINAES. (2013). Instituciones Miembros del SINAES. Retrieved February 26, from http://www.sinaes.ac.cr/universidades_adherentes.htm.

Chapter 14
The Diversity of LIS Programs in Southeast Asia

Shizuko Miyahara

14.1 Introduction

Library and Information Science (LIS) education in Southeast Asia is typical of LIS education in developing countries. These countries have always faced similar difficulties in providing LIS education. Shortages of qualified faculty members, educational facilities, and administrative funds have caused a decrease in the quality of education. These problems are evident in most developing countries.

As stated in Chap. 3, discussions of quality assurance in higher education in Southeast Asia have already begun in the context of the worldwide expansion of higher education. These trends have influenced the field of LIS education and prompted a vigorous discussion. During the past decade, presentations at LIS conferences and workshops such as the Congress of Southeast Asian Librarians (CONSAL) and the Asia-Pacific Conference on Library and Information Education and Practice (A-LIEP) have highlighted the issue of regional accreditation of LIS education. Several Asian scholars have insisted that there is a need to establish an accreditation system based on the US model. In fact, some researchers proposed a regional accreditation scheme in 2002. However, no plan to establish regional quality assurance in Asia has yet been formulated. There are several reasons for this, but the most significant factor seems to be the diversity of LIS education in the region.

Hence, this chapter provides a comprehensive, panoramic view of the historical development of LIS education in Southeast Asia, and examines differences in LIS education in the region. First, it examines the historical development and characteristics of LIS education in Southeast Asia. Second, it reviews previous discussions and studies of LIS quality assurance. It concludes with a discussion of the significance of regional cooperation in Southeast Asia and the possibilities for achieving this.

S. Miyahara (✉)
Sagami Women's University, Sagamihara City, Kanagawa, Japan
e-mail: miyahara_shizuko@isc.sagami-wu.ac.jp

M. Miwa and S. Miyahara (eds.), *Quality Assurance in LIS Education:*
An International and Comparative Study, DOI 10.1007/978-1-4614-6495-2_14,
© Springer Science+Business Media New York 2015

14.2   The Origin of LIS Education in Southeast Asia

Regarding the historical characteristics of Asian higher education, Altbach and
Selvaratnam (1989) note:

> Contemporary Asian higher education is fundamentally influenced by its historical tradi-
> tions. No Asian university is truly Asian in origin – all are based on European academic
> models and traditions, in many cases imposed by colonial rulers, and in others (e.g., Japan
> and Thailand) on voluntarily adopted Western Models.

In the same manner, the origins of LIS programs in the Third World are rooted in
their colonial history. To that extent, the differences in LIS education in the region
began in the colonial era. This section examines the history of LIS education in
Southeast Asia and the relevance of the former colonial powers, the USA and the UK.

14.2.1   US Influence

Most LIS education programs in Southeast Asia were introduced after World War
II. Only the Philippines introduced a library education program before the war. The
US colonial government in the Philippines introduced a training program for native
librarians together with a public education system. In 1914, The University of the
Philippines (UP) offered the first courses in library economy, taught by American
pioneers dispatched from the American Library Association. After 2 years, a 4-year
Bachelor of Science in Library Science (BSLS) curriculum was introduced in the
College of Liberal Arts of the UP Diliman in 1916. This college offered the degrees of
Bachelor of Science in Library Science, Bachelor of Arts, and Bachelor of Philosophy.
It was the first library course at the tertiary educational level in Southeast Asia.

After World War II, the USA contributed to the development of LIS education,
not only in the Philippines but also in Thailand, Indonesia, Japan, Korea, and
Taiwan. Lomrer and Jackson (1959) categorized those Asian countries into three
groups (Table 14.1).

The first countries to receive assistance for libraries were Japan and the
Philippines. Both countries had a long history of library education before receiving
aid from the USA, with both formal and informal library education being conducted.
The second group is Thailand, Indonesia, and Taiwan. There were a few libraries
before US assistance was received but their collections were limited and uncata-
loged, and there was no LIS education. The third group includes Fiji, Burma, Laos,
and Cambodia, where there were few trained librarians or professional schools.

	Period	Countries
Table 14.1 Library assistance in Asia from the USA	Early 1950s	First group (The Philippines, Japan)
	Late 1950s	Second group (Thailand, Indonesia, Taiwan)
	1960s	Third group (Fiji, Burma, Laos, Cambodia)

Below are brief histories of LIS education in Southeast Asian countries. Library education in Thailand began as a part-time subject in Chulalongkorn University in Bangkok. Five American professionals stayed in Thailand from 1951 to 1955 with the support of the Fulbrlight Foundation and certified 77 people. This was called the "Fulbright course." Chualongkorn University established the Department of Library Science in the Faculty of Human Studies and awarded diplomas to students who completed a 1-year course. It offered this course from 1955 to 1967, and 234 students received diplomas. Chulalongkorn University expanded the course to create three programs. Although the diploma program was discontinued, a bachelor's program was offered in the Department of Human Education in 1960 and the graduate school was opened in 1964. Because the USA supported library education, Thammasat, Ramkhamhaeng, and Chiang Mai Universities offered professional programs and Ban Som Dej Teachers College offered a subprofessional program.

The beginning of LIS education was in the 1950s, following independence in 1945. Indonesian LIS education was under the strong control of the colonial Dutch system. The first LIS course was established in Jakarta and was managed by the Ministry of Education and Culture in 1952. This was modeled on the Dutch system, and was a 2-year part-time program intended to educate librarians working for the government. The length of the program was extended to 3 years, and the government expanded it to create a library school. This library school was transferred to the Faculty of Teacher Training and Education Science of the University of Indonesia in 1961, and it became the Department of Library Science in 1963. The university also established a 2-year graduate school program in 1964 and accepted nine students. After this program was discontinued in 1965, the university discontinued its bachelor's program and reestablished the graduate school program.

Short programs are accepted more easily in the Indonesian library system. Gajah Madah University in Jogjakarta and the Institute of Teacher Training and Educational Science in Bandung have offered 5- or 6-month courses since 1968. In Jakarta, the Library Institute of the Department of Education and Culture offers a 1-year part-time program, and the Indonesian Scientific Documentation Center offers a program in special librarianship.

Under the US model, university LIS programs were first offered at the undergraduate diploma or certificate level, and subsequently developed into bachelor's and master's programs. The LIS educational models of Indonesia, the Philippines, and Thailand were based on the US model, whereby universities conferred a bachelor's or master's degree on students. The Philippines has an especially high number of degree programs, far exceeding that of other countries. However, this includes programs in teachers colleges and institutes of education, which have begun to offer training programs for teacher librarians.

The most remarkable achievement of the USA was to establish library schools in various areas and create functional institutions that produced many LIS professionals. US administrators began by establishing institutions such as library schools, preparing LIS programs, student and teacher exchanges, and supporting scholarship. The export of American models of LIS education was an important part of the US library assistance policy and part of an overall plan of educational exchanges.

US library assistance enabled Asian countries to offer LIS education and establish library schools. Most programs began as library science courses to train librarians for public and academic libraries. In the 1970s, Asian countries assumed the management of LIS education because the USA ceased its direct assistance (Bernardo, 1932; Damaso, 1966; Wijasuriya, Lim, Nadarajah, 1975).

14.2.2   The UK Influence

A noticeable difference between the US and UK influences on LIS education in Asia is the recognition of professions and the roles of professional associations. In the USA, universities offer professional programs and grant professional degrees. Professional recognition is based on the possession of a professional degree. Hence, the first goal of aid for library education in Asia was to build local library schools in each country.

In the UK, however, it was not until the 1970s that universities offered formal LIS programs. Traditionally, the UK professional education model placed great value on certificates issued by an appropriate professional association. Unlike the US model, college education was complementary to professional examinations for certificates. Until the 1960s, college lectures in the LIS field were mainly intended to prepare students for library association examinations.

Therefore, formal LIS education in the British Commonwealth often began as postgraduate diploma programs following the UK model, which were later upgraded to master's programs with the addition of a master's thesis.

Singapore and Malaysia also took over the British model. In the first half of the 1980s, there was no degree program in either county, because higher educational institutions did not offer formal LIS programs. Only library associations offered lectures for librarians in diploma/postgraduate diploma programs. In addition, until 1970, overseas candidates had taken the external examination of the Library Association of the UK, following the UK educational model.

14.3   The Development of LIS Education Over Three Decades

As stated above, library educational programs were influenced by US or UK models in Southeast Asia. However, the situation changed after the early 1970s, because the USA ceased to offer generous overseas aid for libraries and the UK abolished the external examinations of the Library Association. This was the second turning point for library educational history in Southeast Asia.

After that, some countries changed their educational systems based on these models, while others maintained the accepted educational model. In addition, curricula for LIS education were transformed by the development of ICT technology and globalization. Asian library schools adapted their educational systems accordingly.

Table 14.2   The number of LIS degrees/certificates in Southeast Asia in the 1980s (Fang, & Nauta, 1985)

Country (number of institutions)	Certificate	Diploma	Bachelor's	Postgraduate diploma	Master's	Ph.D.
Indonesia (5)	3	2	1	0	0	0
Malaysia (1)	0	1	0	1	0	0
Philippines (10)	1	0	11	1	6	0
Singapore (1)	0	0	0	1	0	0
Thailand (7)	0	0	7	0	2	0
Total	4	3	19	3	8	0

This section examines LIS educational development and transformation in the 30 years from 1980, using a guidebook on library schools issued by the IFLA and related studies. The guidebook was published as the *International guide to library and information science education in* 1985. It was revised and republished 10 years later as the *World guide to library, archive and information science education* in 1995. The latest edition is W*orld guide to library, archive and information science education* published in 2007. It focuses on programs related to library science, such as library science, librarianship, library and information science, and information studies, while archive programs have been removed.

14.3.1   LIS Education in the 1970s and 1980s

As stated above, LIS education in Southeast Asia has completely accepted the Western model, relying on funds and professionally trained staff. People could receive advanced education in Western countries during the 1950s and 1960s.

After their independence, most Asian countries were financially unstable; therefore, they did not prioritize library development in national development plans or cultural policy. After the withdrawal of overseas assistance, they could not afford to maintain the Western educational level, hire excellent faculty members with degrees, establish new library schools, or review LIS programs with public funds because of their limited national budgets.

This meant there was a huge gap between the standards of professional education from the West and that of Asian librarians, and it made Asian librarians notice the need for local LIS education. For instance, the Second CONSAL (Congress of Southeast Asian Librarians) held in Manila in 1970 focused on Asian librarianship under the theme of "Education and Training for Librarianship."

According to the 1985 IFLA guidebook, LIS programs were offered at higher educational institutions in five countries: Indonesia, Malaysia, the Philippines, Singapore, and Thailand (Table 14.2). In the US-influenced countries—Indonesia, the Philippines, and Thailand—degree programs in LIS education were already being offered. On the other hand, in Malaysia and Singapore, which were influenced by the UK model, only diploma or postgraduate diploma programs were offered.

It is characteristic of the US model that it gives weight to education at university, and library associations do not offer such education directly. Hence, the number of degree programs in the countries under the US model exceeds that in countries under the UK model.

In both Malaysia and Singapore, library associations played an important role in offering training courses and certification. Academic programs prepared students for the LA examination.

In the late 1970s and 1980s, many information science courses were added to LIS education, and followed the LIS curricula in Western library schools. These included courses on library automation, application of information technology, information retrieval, and online searching in bibliographic databases.

As shown in Table 14.3, this phenomenon occurred in the early 1970s in the University of Indonesia, the MARA Institute of Technology (Malaysia), the University of the Philippines, and in Chulalongkorn University, Sri Nakhaharinwirot University, and Thammasat University in Thailand. Except for the Mara Institute of Technology program, these programs included information science courses as elective rather than core courses. Initially, the information science courses were recognized as part of library science programs. Fundamental curriculum and organizational reform was undertaken in the 1990s (Picache, 1983).

14.3.2   LIS Education in the 1990s

Compared with the 1985, the total number of institutions offering LIS programs increased in the 1990s, especially bachelor's and master's programs. Singapore, which had not previously offered a degree program, established a new LIS graduate program as part of an information science program (Table 14.4).

The number of categories of LIS programs has expanded. Strictly speaking, an LIS program should be defined as an academic program that belongs to an independent department and offers a professional degree such as Bachelor or Master of Library and Information Science (BLIS and MLIS, respectively). However, in actuality, several departments have LIS programs. Nevertheless, LIS programs have diversified. What complicates the creation of a regional accreditation system is the diversity of LIS programs in Southeast Asia.

The situation of the Philippines provides an example. As Tables 14.5 and 14.6 show, the number of institutions in the Philippines is greater than that in any other country. However, the titles of degrees are diverse (see Tables 14.5 and 14.6). Historically, the LIS programs began at the state university, University of the Philippines. Since the government of the Philippines adopted a laissez faire system for private higher education, many private universities have offered library courses in different departments. As a result, a range of degrees is associated with LIS programs.

The concept of LIS has been transformed. The most notable issue was the changes of faculty or department names from "library science" to "library and information science." Courses on Internet technologies and the World Wide Web were

Table 14.3 Information science courses offered in library schools in the 1970s and 1980s (Picache, 1983, p. 46)

University	Program level	Course title	Elective/regular (E/R)	Year introduced
University of Indonesia, Department of Library Science	Master's degree	Introduction to Documentation	E	1970 (?)
	Master of arts	Introduction to Data	E	1978
		Abstracting	E	
MARA Institute of Technology (Malaysia), School of Library and Information Science	Diploma	Fundamentals: Information Storage & Retrieval Systems	R	1975
		Indexing of Information Resources	R	1975
		Information Storage & Retrieval Systems	R	1975
		Computers & Libraries I	R	1979
		Computers & Libraries II	E	1979
		Fundamental of Information Science	R	1979
		Problem Study in Information Science	E	1979
	Postgraduate diploma	Information Storage and Retrieval	R	1979
		Computers & Libraries	R	1979
University of the Philippines, Institute of Library Science	Master of library science	Introduction to Information Science	E	1971
		Information Analysis	E	1980
		Information Technology I	E	1980
		Information Technology II	E	1981
		Information Society	E	1981
		Information Systems Planning & Development	E	1981
		Seminar on Classification & Indexing	E	1981
		Media Technology in Library and Information Science	E	1982
		Quantitative Methods in Library and Information Science	E	1982
Chulalongkorn University	Bachelor's degree BA (LS)	Indexing and Abstracting	E	1971
		Mechanization in Libraries	E	1974
	Master's MA (LS)	Introduction to Information Science	R	1971
Sri Nakhaharinwirot University	Master's MEd (LS)	Information Storage and Retrieval	E	1974
		Systems Analysis in Library Service	E	1974
		Introduction to Information Science	E	1979
		Library Automation	E	1979
		Quantitative Methods	E	1979
Thammasat University	Bachelor's degree BA (LS)	Introduction to Information Science	E	1977

Table 14.4 The number of LIS degree/certificates in Southeast Asia in 1990s (Fang, Stueart, & Tuamsuk, 1995)

Country (number of institutions)	Certificate	Diploma	Bachelor's	Postgraduate/ graduate diploma	Master's	Ph.D.
Indonesia (4)	1	2	0	0	1	0
Malaysia (1)	0	1	0	1	0	0
Philippines (10)	1	0	12	0	5	0
Singapore (2)	0	0	0	1	1	0
Thailand (11)	0	0	10	1	10	0
Total	2	3	22	3	17	0

Table 14.5 LIS degree titles in the Philippines (bachelor's Vallejo, 1996, p. 80

Title of degree	Number of institutions
Bachelor of Library Science: BLS	2
Bachelor of Arts (major in library science): AB	3
Bachelor of Science in Education (major or with specialization in library science): BSE	27
Bachelor of Science in Elementary Education (with specialization in library science): BSE ED	11
Bachelor of Science in Library Science: BSLS	1

Table 14.6 LIS degree titles in the Philippines (master's Vallejo, 1996, p. 80)

Title of degree	Number of institutions
Master of Arts in Library Science: MALS	8
Master of Library Science: MLS	2
Master of Science in Library Science : MSLS	1
Master of Arts in Education (with specialization in library science): MAEd	1
Master of Arts in Teaching (with specialization in library science): MATLS	1
Master of Education (with specialization in library science): MEd LS	3

added in the 1990s. These emphasized "information science," reflecting market demand for ICT information handling skills, and shifted the core of these programs from traditional library science to ICT and knowledge-oriented informatics.

14.3.3   LIS Education in 2000s

In the 2000s, the expansion of higher education in Asia had a noticeable effect on LIS education.

The number of graduate programs with master's degrees also increased. One remarkable aspect of LIS programs in the 2000s is the increasing number of graduate programs. All five countries have commenced LIS graduate programs.

Table 14.7 The number of LIS degrees/certificates in Southeast Asia in the 2000s (Axel, 2007)

Country	Certificate	Diploma	Bachelor's	Postgraduate/ graduate diploma	Master's	Ph.D.
Indonesia (12)	0	6	6	0	3	1
Malaysia (3)	0	0	4	1	5	2
Philippines (37)	1	1	30	0	20	0
Singapore (1)	0	0	0	0	1	0
Thailand (17)	0	0	14	1	10	0
Total	1	7	54	2	39	3

Table 14.8 Title of doctoral degree in Southeast Asia (Axel, 2007)

Country	University	Title
Indonesia	Universitas Gadjah Mada	Doctor
Malaysia	International Islamic University	Doctor of Philosophy in Library and Information Science
	MARA Institute of Technology	Doctor of Philosophy (Information Management)

In particular, two countries began doctoral programs in Indonesia and Malaysia (Tables 14.7 and 14.8).

However, mainstream LIS degrees are not concentrated in master's programs as they are in the USA. The academic structure of LIS programs has been expanded by the addition of higher degree programs. This trend is in response to the job market for LIS graduates.

The requirement for librarians has also increased. In most Asian countries, the requirement to hire a professional librarian limited the number of qualified applicants. There is a strong connection between the professional education at bachelor's/master's level and job requirements. For this reason, the influence of the US model, which focuses on professional university education and degrees, has been expanding in Southeast Asia.

14.4   Discussion of Quality Assurance in LIS Education

As of 2013, there is no regional accreditation system for higher educational programs in Asia. Moreover, no Asian country has an accreditation system for LIS programs that is administered by a library association like that of the USA. Instead of program accreditation, accreditation at the institutional level is undertaken in some countries. To this extent, LIS programs can be said to be accredited at the institutional level.

More importantly, the institutional accreditation system itself remains a problem. The mechanisms of institutional accreditation systems in Asian countries are diverse.

Moreover, it has been pointed out that an indirect system of quality assurance has been established by governments in various countries. Some countries, such as Myanmar and Cambodia, have no accreditation system. The diversity of institutional accreditation systems makes it difficult for Asian universities to assure the same quality of education at the regional level. Because LIS programs in Southeast Asia are also diverse, a common accreditation system at the program level is required to assure the same quality of professionalism across national borders.

14.4.1   Global Research and Discussions

The issue of regional accreditation of LIS professional programs has been discussed in regional LIS education conferences and workshops for a number of years.

As noted in Chap. 11, the study of quality assurance in higher education has become an active area of research. This section describes the recent trend in previous studies to investigate quality assurance in LIS education.

The education and training section of the IFLA undertook an initial global study of quality assurance systems for LIS education. Since 1977, the Education and Training section investigated mutual recognition of LIS programs, and presented the research outcomes at international conferences. Comprehensive studies of global quality assurance systems were presented by Broady-Preston and Harriaon (2002), Kigongo-Bukenya (2005), Tammaro (2005), and Weech and Tammaro (2007). The studies focused on the LIS programs of the USA, the UK, Sweden, and Australia. Research on regional cooperation in quality assurance investigated issues in Europe.

Researchers from non-Western countries presented research about proposals for or information about the current status of quality assurance systems in their own countries or areas. This research covered countries such as India (Sarkhel, 2006, 427–431), Pakistan (Ameen, 2007), and Malaysia (Singh, 2007, 87–112). The studies or proposals for regional cooperation to establish a regional quality assurance model are for East Asia (Lin, 2004), Lin and Wang (2006), Southeast Asia (Shaheen, 2003, 58–69), (Khoo et al., 2003, 131–149), (Chaudhry, 2007, 25–31), the Asia-Pacific region (Abdullahi, & Kaur, 2007, 56–69), the Middle East (Rehman, 2008, 366–382), (Rehman, 2007, 70–86), and Africa (Ocholla and Bothma, 2007, 55–78).

In Japan, studies of quality assurance systems in LIS education are very few, and include those of Ohshiro (1998), Sakai (2002), and Mizoue (2004). These studies focus on the introduction of the US accreditation system. Recently, studies of a global accreditation model and the contribution of Japanese scholars have been presented by Miwa et al. (2006), and Miwa, Kasai, and Miyahara (2007).

14.4.2   Discussion of Southeast Asia

The initial discussion of the accreditation of LIS programs and degrees in the Southeast Asian region was undertaken at the International Conference for Library and Information Science Educators in the Asia Pacific Region held in Kuala Lumpur in 2001. The conference surveyed LIS education programs in the region, discussed the core competencies of the profession and examined some issues. As a background to the discussion, Chaudhry (2007) pointed out that an accreditation scheme for the region would be useful in enhancing the quality and acceptability of LIS degrees, and would provide a greater flexibility for the mobility of library and information professionals in the region.

In 2003, at CONSAL XII in Brunei, the issues of accreditation and certification were raised and a proposal for the development of a regional accreditation and certification scheme was included in the conference resolutions. At the A-LIEP conference in 2007, a whole session was devoted to accreditation issues.

In 2000, Majid, Chaudhry, Foo, and Logan (2002) conducted a questionnaire survey of LIS schools in Southeast Asia concerning a regional accreditation scheme. It was clear that most of the schools surveyed agreed on the need for the accreditation of LIS degrees in the region, and expressed interest in participating in an accreditation scheme. In 2002, Majid et al. proposed a model of quality assurance in Southeast Asia. They undertook another questionnaire survey of LIS schools in the region to gather views on various issues related to a regional accreditation scheme, including the need for such a scheme, the geographical scope of accreditation, and coordination, duration, cost, and potential problems.

Rehmen (2007) and Koo, as well as Majod and Chaudry (2003) also discussed issues of the regional accreditation system. They identified an appropriate regional body to administer the accreditation system, the problem of determining a framework and formulating a set of standards.

14.5   Conclusion

This article takes a comprehensive and panoramic historical view of LIS programs in Southeast Asia. The diversity of Southeast Asia is a characteristic that attracts many people. However, this diversity is an obstacle to building common regional systems such as an accreditation system. As a result, although it is more than 10 years since it was initially proposed at a conference, no significant progress has been made. It seems that the diverse programs and different situations of LIS education make progress slow.

This study clarifies that LIS programs in Southeast Asia are rooted in colonial systems, and there is a division between systems that originated with the US and UK models. These models are quite different in terms of professional education and recognition. Professional recognition under the UK model was based on certification by the Library Association. The association was involved in recognition and direct examination of candidates.

On the other hand, the US model emphasized professional education at institutions. Academic degrees from accredited programs are the only acceptable qualification under that model and a form of license is required to work as a library professional. Therefore, the accreditation system for LIS degree programs was developed at an early stage to ensure and maintain the quality of professionals.

In the 1970s and 1980s, library science curricula were expanded to include information science, with the addition of courses on library automation, application of information technology, information retrieval and online searching in bibliographic databases. In the 1990s, courses on Internet technologies and the World Wide Web were added. Many universities renamed their departments of library science to add "information" (Khoo, Majid, Lin, 2009, 196-198).

In Southeast Asia, the number of LIS programs has expanded since the 1990s in the wake of a boom in higher education resulting from economic growth. In addition, changes in higher educational policy make universities change from institutions that train a limited number of elites to ones that provide educational opportunities for more students. In the past two decades, the massification of higher education has progressed and affected the development of LIS education. As noted above, the number of institutions that offer LIS programs has increased, and the content of LIS programs has diversified. This diversity makes it difficult to establish a common accreditation system at the regional level.

As globalization progresses, the mobility of people accelerates. In the library field, the number of librarians, LIS students and LIS scholars seeking jobs overseas is expected to increase. The main advantage of a regional accreditation scheme is that it assures students and employers of the quality of a program and improves the mobility of graduates. We should pay close attention to further discussions about a regional quality assurance system.

References

Abdullahi, I., & Kaur, T. (2007). The importance of accreditation of LIS program in Asia and Pacific region. In C. P. Lin et al. (Eds.), *Proceedings of the Asia-Pacific Conference on Library and Information Education and Practice 2007 (2nd A-LIEP 2007)*. Taipei: Department/Graduate Program of Information and Communications, Shih-Hsin University.

Altbach, P. G., & Selvaratnam, V. (Eds.). (1989). *From dependence to autonomy: The development of Asian universities*. Dordrecht: Kluwer.

Ameen, K. (2007). Issues of quality assurance (QA) in LIS higher education in Pakistan. *73rd IFLA General Conference and Council, Durban, South Africa*, August 19–23, 2007. http://www.ifla.queenslibrary.org/IV/ifla73/papers/114-Ameen-en.pdf.

Axel, S. (Ed.). (2007). *World guide to library, archive and information science education* (3 new and completely revth ed.). Munich: K.G. Saur.

Bernardo, G. A. (1932). *The status of the popular library movement in the Philippines*. Quezon City: University of the Philippines.

Broady-Preston, J., & Harriaon, J. (2002). The quality assurance agency (QAA) and subject review: The viewpoint of the assessor. *Libri, 52*(4), 195–198.

Chaudhry, A. S. (2007). Collaboration in LIS education in Southeast Asia. *New Library World, 108*(1/2), 25–31.

Damaso, C. (1966). Library education in the Philippines. *Journal of Education for Librarianship, 6*(4), 96–104.

Enser, P. (2002). The role of professional body accreditation in library & information science education in the UK. *Libri, 52*, 214–219.

Fang, R. J., & Nauta, P. (Eds.). (1985). *International guide to library and information science education: A reference source for educational programs in the information fields world-wide*. Munich: KG Saur.

Fang, R. J., Stueart, R. D., & Tuamsuk, K. (Eds.). (1995). *World guide to library, archive and information science education* (2nd ed.). Munich: KG Saur.

Khoo, C., Majid, S., & Chaudhry, A. S. (2003). Developing an accreditation system for LIS professional education programs in Southeast Asia: Issues and perspectives. *Malaysian Journal of Library & Information Science, 8*(2), 131–149.

Khoo, C., Majid, S., & Lin, C. P. (2009). Asia: LIS Education. In I. Abdullahi (Ed.), *Global library and information science: A textbook for students and educators*. Munchen: K. G. Saur.

Kigongo-Bukenya, I. M. N. (2005). Global library and information (LIS) concerns: The case for international accreditation of LIS. *71th IFLA General Conference and Council, Oslo, Norway*, August 14–18. http://www.ifla.queenslibrary.org/IV/ifla71/papers/181e-Kigongo-Bukenya.pdf.

Lin, C. P. (2004). *The challenges and opportunities of regional cooperation in LIS education in East Asia*. Retrieved March 14, 2010, from http://www.ifla.org/IV/ifla70/papers/065e-Lin.pdf.

Lin, C. P., & Wang M.-L. (2006). *Regional LIS Education Cooperation in Asia, a continuing effort*. http://nccur.lib.nccu.edu.tw/bitstream/140.119/21882/1/107-LinWang-en.pdf.

Lomrer, A., & Jackson, W. V. (1959). Education and training of librarians in Asia, the Near East, and Latin America. *Library Trends, 8*, 2.

Majid, S., Chaudhry, A. S., Foo, S., & Logan, E. (2002). Accreditation of library and information studies programmes in Southeast Asia: A proposed model. *Singapore Journal of Library and Information Management, 32*, 58–69.

Mizoue, C. (2004). America no toshokan jyohogaku kyoiku to ninsho hyoka. [Library schools and accreditation system in the US]. *Library, Information and Media Studies, 2*(2), 33–44. In Japanese.

Miwa, M., Kasai, Y., & Miyahara, S. (2007). Toward mutual accreditation and creit exchange among LIS programs in Asia-Pacific Region. In C. P. Lin et al. (Eds.), *Proceedings of the Asia-Pacific Conference on Library and Information Education and Practice 2007 (2nd A-LIEP 2007)*. Department/Graduate Program of Information and Communications, Shih-Hsin University: Taipei.

Miwa, M., Ueda, S., Nemoto, A., Oda, M., Nagata, H., Horikawa, T. (2006). Final results of the LIPER project in Japan. *72nd IFLA Council and General Conference*, Seoul, Korea, 20–24 August. Retrieved January 25, 2013, from http://web.keio.jp/~uedas/papers/107-Miwa-en.pdf.

Ocholla, D., & Bothma, T. (2007). Trends, challenges and opportunities for LIS education and training in Eastern and Southern Africa. *New Library World, 108*(1/2), 55–78.

Ohshiro, Z. (1998). America gashukoku no tosyokan jyohogaku niokeru nintei. [Accreditation in library and information science education in the United States. *The Library World, 50*(4), 168–177. In Japanese.

Picache, U. G. (1983) *Education of library and information personnel in the ASEAN region*, CONSAL VI, The library in the information revolution: Proceedings of the Sixth Congress of Southeast Asian Librarians, Singapore, 30 May–3 June, Maruzen.

Rehman, S. (2007). Accreditation of LIS programs in the Arabian Gulf Region. In C. P. Lin et al. (Eds.), *Proceedings of the Asia-Pacific Conference on Library and Information Education and Practice 2007 (2nd A-LIEP 2007)*. Taipei: Department/Graduate Program of Information and Communications, Shih-Hsin University.

Sakai, Y. (2002). Hokubei no toshokan jyohogaku kyoiku no genkyo.[Library and information science education in North America: Expanding the programs for information professionals in broader environments]. *The Journal of Information Science and Technology Association, 52*(7), 354–363. In Japanese.

Sarkhel, J. K. (2006). Quality assurance and accreditation of LIS education in Indian Universities: Issues and perspectives. In Khoo et al. (Eds.), *Proceedings of the Asia-Pacific conference on library and information education and practice 2006*. Singapore: School of Communication & Information, Nanyang Technological University.

Sangpichitara, U. (1979). *The development of the modern library and library education in Thailand*. Ph.D. thesis, University of Michigan, USA.

Shaheen M. et al. (2003) Accreditation of library and information studies programmes in Southeast Asia: A proposed mode. *Singapore Journal of Library and Information Management*, Vol.32, 58–69.

Singh, D. (2007). Accreditation of library and information science education programs: The Malaysian experience. In C. P. Lin et al. (Eds.), *Proceedings of the Asia-Pacific Conference on Library and Information Education and Practice 2007 (2nd A-LIEP 2007)*. Taipei: Department/ Graduate Program of Information and Communications. Shih-Hsin University.

Tammaro, A. M. (2005). Recognition and quality assurance in LIS: New approaches for lifelong learning in Europe. *Performance Measurement and Metrics, 6*(2), 67–79.

Vallejo, R. (1996). Library and information science education in the Philippines: A country report. In K. M. Chavalit (Ed.), *Introduction to ASEAN librarianship: Library education and training*. ASEAN Committee on Culture and Information: Bangkok.

Weech, T. L., & Tammaro, A. M. (2007). *Feasibility of international guidelines for equivalency and reciprocity of qualifications for LIS professionals: Progress report.* http://www.ifla.org/ VII/s23/index.htm.

Wijasuriya, D. E. K., Lim, H. T., & Nadarajah, R. (1975). *The Barefoot Librarian: Library developments in Southeast Asia with special reference to Malaysia*. Hamden, CT: Linnet Books.

Chapter 15
Regional Quality Assurance System: Gulf Cooperation Council (GCC) Member Nations in Middle East

Sajjad ur Rehman

15.1 Context

The six nations of the Gulf Cooperation Council (GCC) have similarities in linguistic, socio-politico-economic, and cultural domains. Member nations include United Arab Emirates, Bahrain, Saudi Arabia, Oman, Qatar,, and Kuwait. GCC was established in 1981, the original Council comprised the 630-million-acre (2,500,000 km) Persian Gulf states. The unified economic agreement of 1981 has been the integrative imperative. The forum has regulated in various fields such as religion, education, defense, finance, trade, customs, tourism, legislation, immigrant labor, and administration. These nations have also strived for cooperation in the domains of science, technology, mining, agriculture, water, and animal resources. These policies are aimed at strengthening ties between their peoples. They wish to have common currency, market, and human trafficking policies. Development of most economies heavily relies of immigrants who even outnumber the local populace in many member nations. This area has some of the fastest growing economies in world, mostly due to a boom in oil and natural gas revenues coupled with a building and investment boom backed by decades of saved revenues. National and regional funds have several hundreds of billions dollars of assets. GC also maintains a patent office. A GCC common market was launched in 2008. The common market grants national treatment to all GCC firms and citizens. In 2009, Bahrain, Kuwait, Qatar, and Saudi Arabia announced creation of Monetary Council, a step toward establishing a shared currency. All these nations have monarchs with a tight political climate. Kuwait has a regulated parliament, but the monarch controls government.

S. ur Rehman (✉)
Kuwait University, Jamal Abdul Nasser St, Kuwait
e-mail: rehman@gmail.com

M. Miwa and S. Miyahara (eds.), *Quality Assurance in LIS Education:*
An International and Comparative Study, DOI 10.1007/978-1-4614-6495-2_15,
© Springer Science+Business Media New York 2015

15.2   Library and Information Education in the GCC

In this section, we give a basic profile of library and information science (LIS) education in the GCC nations. In this LIS education community, the program founded at King Abdulaziz University, Saudi Arabia (KAU) in 1973 is the oldest in the region and Master's program recently developed at the American University for Emirates (AUE), developed in 2011/2012, is the latest. The undergraduate program at PAAET (Kuwait) was established in 1977. It seems that the period of 1984–1988 was the most fertile for establishing new programs in this region as three Saudi programs at University of Al-Ummal Qura's (UQ) Men and Women wings (Mecca) and King Saud University (KSU) were established during this period. Also the undergraduate programs of Sultan Qaboos University SQU (Muscat, Oman) and Qatar University (QU) were founded in 1986 and 1988 respectively. MLIS at Kuwait University (KU) started functioning in 1996. Two year back, in 2011, Qatar University's program was phased out.

Two significant developments are not factored in the analysis of information for this chapter and its write-up. These are: (1) phasing out of QU undergraduate program, and (2) beginning of a graduate program of LIS in Emirates (AUE).

Gathered information about the placement of LIS programs in the GCC higher education institutions. There appeared to be two leading choices for the placement of these programs in different colleges. Three of these programs were located in the College of Social Sciences, while two others were found to be part the College of Arts. SQU program was located in the College of Arts and Social Sciences while QU was placed in the College of Arts and Science. PAEET was the only program situated in the College of Education. It means that eight degree awarding LIS programs in six nations had different placement arrangements.

One distinctive feature of LIS programs in GCC nations is that seven of the eight are predominantly undergraduate. It is only KU that does not have an undergraduate major though it has been offering a minor in information studies. Among these programs, KAU is the only one that offers 3-degree programs of bachelor's, master's, and that of Ph.D. The graduate programs of this university had 24 and 11 students in one academic year that were all pursuing research track in their degree programs, meaning there is no structured coursework required in their graduate degree programs. It is worth noting that majors in the undergraduate programs in both men and women wings of UQU and the minor of KU have an orientation toward information science or studies. SQU and KAU have recently adopted this nomenclature during the last 3–4 years. These programs are not designed to cater for the traditional markets of LIS programs. Master's program at KU and AUE are typical programs patterned on the model of ALA-accredited programs with a coursework structure of 36 credit hours. AUE also requires thesis research as a graduation requirement. KU MLIS program has had no thesis provision as yet. This program made two attempts back in 2010 and 2012 for adopting the identity of information studies; though these efforts have been inconclusive. All undergraduate programs in the region are patterned on semester system, requiring credit hours for major, minor, and other segments. The two Master degree programs at KU and KAU have 20 and 50 students respectively with an annual intake of 40–50 students at KU. Averagely, KU graduates about 20 students every year, indicating a dropout of 50–60 %.

It is worth-noting that most of these undergraduate programs were designed on the pattern of Egyptian model, as these universities mostly benefitted from the services of Egyptian consultants. Egyptian experts have left a pioneering mark on the LIS education in this region (Rehman, 2012).

Rehman found that undergraduate programs in GCC universities have large enrolments. UQU had an enrolment of about 1,800 students in both wings of men and women. PAEET, Kuwait also had an enrolment of 700. The number of undergraduate students at QU was 400. Enrolment of undergraduates at KAU, KSU, and SQU was in the range of 200–300.

Bahrain, a small GCC nation, also needs LIS professionals desperately. Young and Ali reported back in 1992 that there were only nine Bahraini Master's degree holders while needs were enormous. In response to these needs, University of Bahrain started a postgraduate diploma in learning and information resource centers and a large number of graduates had been produced in this program. Later the university abandoned this program. Also quite a few bachelor degree holders were hired from overseas and the number of professionals possessing bachelor and Master degree exceeded 50 around early 1990s. Then University of Bahrain decided to send 2–3 bachelor degree holders to earn their Master's degree from the USA or Europe.

Assessed Omani situation of library and information professionals and noted a serious dearth in the Sultanate. He had inventoried 189 libraries in country and noted that an acute shortage of professionals hampered their development. Majority of the staff had no professional qualifications and they heavily relied on expatriate workforce. Against this background, they had established the Department of Library and Documentation in the College of Arts at Sultan Qaboos University. Since then hundreds of graduates have passed out from the SQU of Oman who should have entered job market, changing the employment situation.

United Arab Emirates is a fast growing nation in this region. It has also developed an elaborate IT infrastructure and Dubai Media City and Computer City are attracting attention from all over the world. The growth in higher education sector has been unprecedented in the region and a large number of institutions have been developed. It has created a strong need of information professionals. Against this backdrop, non-availability of a formal degree program in this country made it difficult for employers to manage library and information centers. Had observed that these libraries were understaffed and consequently underutilized. She emphasized that country needed professionals who were well familiar with environmental conditions. She stressed that a formal degree was imperative for the nation. Then Aman and Mika (2004) visited the country and consulted on developing a bachelor's degree program at Abu Dhabi University. It was a detailed blueprint, but it never materialized. Of late, a couple of years back, a new Master's degree program in library and information science has been instituted at the American University in Emirates. It is a 36-credit hour program, having the provision of a thesis. Since it is relatively a new program, no evaluation is yet conducted. This program may produce much-needed professional workforce for United Arab Emirates.

15.3 Market Needs: Global Perspective

Changes in library and information science (LIS) education have been profound, pervasive and universal. During the last couple of decades, this process of change has accelerated as new areas of studies have emerged in the field and many interdisciplinary academic programs have surfaced that include information management, knowledge management, content management, information architecture, digitization, and archival and record systems. Back in 1999, TFPL completed surveys of the market and identified how the field has opened new opportunities in the areas of information and knowledge management. Abell (1998), the principal consultant of TFPL, emphasized that there were many new opportunities for information professionals and if they did not benefit from them, new opportunistic professions may take lead.

The hallmark study of KALIPER (2000) had identified key factors that had prompted new trends in LIS education. These factors included demands of students, employers, graduates, and professional associations for graduate competencies; growth and expense of supporting emerging technology; internal campus relationships and positioning; availability and/or presence of faculty with new subject expertise; competition from other LIS programs; and availability of financial support for innovation. The six trends delineated in the report indicated the LIS schools are increasingly:

1. Addressing broad-based information environments and information problems in curricula
2. Emerging with a distinct core that is predominantly user-centered
3. Increasing the infusion of information technology into their curricula
4. Experimenting with specialization within the curriculum
5. Offering instruction in diverse formats
6. Expanding curricula by offering related degrees at the undergraduate, master's, and doctoral levels

15.4 Assessment Domains and Practices

It is understood that in order to introduce changes in LIS programs, the aforementioned trends and market needs serve as useful and relevant criteria for assessment of LIS programs. Such efforts must be preceded by systematic evaluation of the context, strategies, curriculum, facilities and resources, and other related factors. In this regard, studies of perceptions of students, faculty, alumni, and other stakeholders play an important role. A number of perception-based surveys are found in literature (Blankson-Hemans & Hibberd, 2004; Edomi & Ogbomo, 2001; Zainab, Edzan, & Rahman, 2004; Genoni, Exon, & Farrelly, 2000; Genoni & Smith, 2005; Jefferson & Contreras, 2005; Loughridge & Speight, 1996; Mohai, 1999; Yen, Chen, & Lee, 2003). We can derive the following points of significance from these studies:

1. Changes in the information market are pervasive.
2. LIS education has undergone major changes during the last few decades.

3. New fields of study and areas of practice have influenced the LIS education programs. These have an interdisciplinary nature.
4. Academic programs of LIS are reconfigured in the light of market needs, based on fresh efforts of competency definition and validation.
5. Competencies are defined on the basis of market needs' assessment, demands of the employment market, situation and profile of the academic programs, and percepts of graduates and other stakeholders.
6. The academic programs of LIS need to be rejuvenated and redesigned, based on continuous efforts of strategic planning, implementation, and evaluation.

15.5 Quality Assurance Model

Keeping in view the afore-elaborated discussions, international organization of IFLA has been active. This organization's Education and Training Division (SET) took it upon itself to develop a viable model of quality assurance. For this purpose, this Division engaged Tammaro (2005, 2007), to propose a models of quality assurance. She applied this model in an international survey and reported the findings. She studied evaluation and quality assurance among LIS programs in the European Union. Four types of criteria of inputs, activities, outputs and outcomes were used for this assessment. The study covered the areas of accrediting agency, frequency, and the areas covered. This study is worth replication in other regions with appropriate adjustments. In India, Sarkhel (2006) investigated the role of Indian University Grants Commission in accrediting LIS programs and ensuring quality assurance. The author developed a set of indicators on the basis of an understanding of global developments in the activities and services of libraries and information centers, the national environment, the outcome of National Assessment and Accreditation Council (NAAC). During the last decade, a number of studies were conducted about the evaluation and quality assurance assessment in Thailand, Poland, and Latvia. These studies indicated the value of assessment of education programs and how local contexts warranted adjustments in the use of criteria and role of different agencies in accreditation process (Holma & Pakalna, 2007; Saladyanant, 2006; Wozniczka-Paruzel, 2003).

15.6 LIS Assessment in GCC

In this chapter, we review the assessment of library and information education programs in the GCC region. We rely on the available body of studies. These have been in the realms of evaluation, quality assurance, and accreditation possibilities and prospects.

There is a need that all education programs are evaluated periodically in a systematic and comprehensive manner. Further, such evaluative exercises must lead to logical outcomes for introducing meaningful changes in the areas of strategy, academic policies, and curriculum revision. Another assumption is that the academic programs need to ensure quality assurance so that uniformity is maintained among those programs that share inherent affinities.

The LIS programs introduced in the region during the last three decades are mostly patterned on the academic structure of semester system and by and large these have been conducted at the undergraduate level. Criteria for evaluation of LIS programs should address the vital aspects of:

- Academic content (curriculum and syllabi)
- Student enrollment
- Faculty
- Instructional facilities physical infrastructure)
- Instructional resources such as laboratories and library

Evaluation of an academic program can be done using a variety of approaches. These essentially include:

1. Self-study
2. External evaluation
3. Accreditation

15.6.1   Self-Study

The academic management of an educational program is responsible for conducting self-study. Normally, a committee is entrusted with the task, having clear targets, timeframe, and output. Faculty involvement is essential for the success of such a study. External consultants and accreditation teams also make it mandatory that self-study reports are provided to them before their visit.

15.6.2   External Consultant

In the absence of accreditation, many universities have a mandatory visit of an external consultant periodically. Such consultants are renowned academics. They meet with academic managers, faculty members, and other stakeholders such as students, alumni, and employers. The consultants are required to produce report in the light of specified criteria and guidelines.

15.6.3   Accreditation

ALA, CILIP, and ALIA are the national agencies that have defined criteria and mechanics for accrediting academic programs in North America, the UK, and Australia, respectively. Details of this practice and its ramifications for the GCC region are explicated in a later section of this chapter. As of now, the GCC region has no accreditation system in place.

A few years back, Rehman (2008) conducted a seminal evaluation study of the then existing LIS programs in the GCC region. This study covered both the evaluation strategies of self-study and review by external experts. The perceptions of academic management about accreditation and certification were explored that are reported in a later part of the chapter.

The review was focused on the following aspects:

- Situation of LIS programs in the six GCC nations in regard to their organizational placement, student enrollment, faculty, computing facilities, and educational resources
- Practices of evaluation of the LIS programs using either self-study or external review
- Perceptions of leading academics about certification and accreditation in these programs

Information was collected about the following elements:

- Profile of each program
- Practices of self-study or review by external assessors of these programs
- Outcomes of the evaluation exercise
- Perceptions about accreditation

15.7 Profile

Rehman (2008) presented important features of LIS programs of education in the GCC nations in a tabular form (Table 15.1). This table illustrates the commonalities and differences in some basic features of founding dates, organizational placement in respective colleges, and student teacher ratios of these programs. One significant dimension is the age of these programs, reflecting overall development pattern of LIS education in the region. One distinctive feature of the LIS programs is that seven of the eight are predominantly undergraduate programs. KU and AUE do not have an undergraduate major though KU has been offering a minor in information studies.

Student enrollment and student teacher ratios are vital signs of the health of library and information education. The undergraduate programs in these universities had large enrollments. UQU had an enrollment of about 1,800 students in both the wings of men and women. PAEET, Kuwait also had an enrollment of 700. The number of undergraduate students at QU was 400. Enrollment of undergraduates at KAU, KSU, and SQU was in the range of 200–300.

The largest number of 21 faculty members at PAEET has to be viewed in relation to an enrollment of 700 students, resulting in a student–teacher ratio of 1:33. UQU also had high student–teacher ratios of 1:57 and 1: 73 for men and women wings respectively. Ratios for SQU and QU were also at 1:31 and 1: 44. For the combined strength of graduate and undergraduate students of KAU, the ratio was 1:20. The ratio for KSU was the lowest of 1:11. For KU, separate ratios for graduate and undergraduate students were 1:11 and 1:20 respectively. Evidently five of the eight undergraduate programs had the ratios exceeding 1:30 while two of the five even exceeded 1:50. These ratios are much higher than the norm. The faculty members

Table 15.1 Profile of LIS education in the GCC nations

Institution	Year established	College	Number of students	Number of faculty
KAU	1973	Arts	Bachelor: 242 Master: 24 Ph.D.: 11	Prof.: 3 Assoc. Prof.: 5 Asstt Prof.: 4 Lecturer: 2 Ph.D. students: 5
QU	1988	Arts & Sciences	Bachelor: 400	Prof.: 1 Assoc. Prof.: 3 Asstt Prof.: 5 TAs: 2
SQU	1986	Arts & Social Sciences	Bachelor: 275 Master: 22 Diploma: 8	Prof.: 1 Assoc. Prof.: 2 Asstt Prof.: 7 TAs: 4 Ph.D. students: 2
UQU women	1987	Social Sciences	Bachelor and Media Center Certificate: 850	Assoc. Prof.: 4 Asstt Prof.: 11 TAs: 7 Ph.D. students: 4
KU	1996	Social Sciences	Bachelor minor: 51 Master: 45 One required service course for 450 students and another required service course for 200 students every year	Prof.: 2 Assoc. Prof.: 3 Asstt Prof.: 5 TAs: 4 Ph.D. students: 5
PAAET	1977	Education	Bachelor: 700	Prof.: 1 Assoc. Prof.: 1 Asstt Prof.: 14 Lecturer: 5 TAs: 14 Ph.D. students: 6
KSU	1986	Arts	Bachelor: 200	Prof.: 4 Assoc. Prof.: 5 Asstt Prof.: 9 TAs: 5
UQU men	1984	Social Sciences	Bachelor: 944	Assoc. Prof.: 2 Asstt Prof.: 11 Ph.D. students: 4

cannot provide guidance and individual counseling if the number of students is that high. A shortage of faculty members in these programs also reflects on the quality of education. For accrediting agencies, this is a always a points of concern.

One of the reasons of this shortage is that there is no tradition of teaching and research assistants or tutors in the GCC programs who can assume effective instructional roles for many basic courses. Also there is no tradition of visiting and adjunct faculty who always bring viable practitioner perspectives in the instruction of applied coursework. In the Western LIS programs, these bring a clear strength to the faculty.

15.8 Resources and Facilities

Another important aspect of evaluation is related to resources and facilities. No academic program can perform effectively unless it is equipped with resources and facilities needed for its instruction. GCC nations are affluent and they are expected to have first-rate facilities and resources. Rehman (2008) presented information about computing facilities, electronic classrooms, audiovisual facilities, library resources, and teaching facilities of these academic programs. Tabulated data have been presented in Table 15.2. All the eight programs had computing laboratories with varying extent of facilities. Among those that provided detailed information, UQU Women was reported to have two networked laboratories. The Men wing of the same university had 80 networked workstations. It is worth mentioning that these two programs have the student strength of about 1,800. SQU reported 25 workstations in the laboratory for the student strength of about 300, meaning 1 workstation for about 12 students. Each of the KSU's four laboratories had 30 networked workstations while the number of students was 200, meaning that there was one pc for every 3–4 students. KU's computing laboratory had 15 workstations; one for about three students. There was a dedicated undergraduate computer laboratory while the undergraduate students shared college facilities. They however needed larger laboratories with additional pc units in order to accommodate larger classes. All the laboratories had Internet connections.

Five of the eight programs did not have electronic classrooms. The other three—KU, QU, and UQU Men—had projection facilities and Internet connections in their classrooms. None of them reported that these classrooms were connected with the central media facilities of the university. Three programs did not have audiovisual facilities. KSU reported of having 12 TVs, video equipment, recorders, etc. Other schools had projection facilities. At SQU, a central unit was equipped with learning technology and each college also had a small unit to facilitate local needs. This program had children laboratory and a facility for bibliographic activities.

As far as library resources are concerned, information could not be ascertained about four programs. Since all of these programs save KU's MLIS program use Arabic medium of instruction, it is pertinent to assess the resources both in English and Arabic languages. KSU and QU had subscriptions for 11 and 13 titles; out of which 5 and 3 were Arabic. KAU subscribed to 9 Arabic and 27 English titles. KU had the largest number of subscriptions of 10 Arabic and 80 English titles. When it comes to monographs, the largest collection of seven and eight thousand volumes for Arabic and English titles was available at KAU. The second largest collection was at KU, which had 2,000–3,000 Arabic and English language volumes. Respondent from QU commented that the collection was very poor. PAEET, with a student body of 700, reported the monograph collection of 1,500 and 700 for Arabic and English languages. KSU had a collection of 1,000 and 400 in English and Arabic. The information for periodical and monograph collections is incomplete, yet it indicates that most programs have inadequate resources while the number of students in these institutions is high. If per capita number of periodical subscriptions

Table 15.2 Resources and facilities

Institution		KAU	QU	SQU	UQU women	KU	PAAET	KSU	UQU men
Computer laboratories		X	X	X	X	X	X pcs and printers	X	X
Electronic classrooms		None	X	None	None	X	None	None	X
Audiovisual facilities		None	None	X	None	X	None	X	X
Periodical subscriptions	Arabic	9	3	No info.	No info.	10	No info.	5	No info.
	English	27	10	No info.	No info.	80	No info.	6	No info.
Monograph collection	Arabic	7,000	Very poor	No info.	No info.	2,000	1,500	1,000	No info.
	English	8,000	Very poor	No info.	No info.	3,000	700	400	No info.
Automation package		Horizon	Unicorn	Afaac (locally developed); library converting to Unicorn	Horizon; Dspace digital library system	Horizon	None	None	Horizon

and monograph collections is computed, it will not present an encouraging scenario. It is worth exploring what factors are responsible for this apparent weakness in these oil-rich nations.

Two points have to be made. First, Arabic is the medium of instruction in all undergraduate programs. Yet, number of textbooks in Arabic is far and few. LIS periodicals in Arabic are also numbered and these are not regarded worthwhile in their scholastic standing and repute. Most Arabic books are translated version of those English books that were popular in 1980s and 1990s. These are serious short-comings and there is no easy shortcut to remedy this situation. Students have minimal skills in English language and they have to use whatever meager resources are available in Arabic.

Secondly, since full text periodical repositories present a workable solution, many of these programs can benefit from this resource. Again, there is an issue of the language of the e-resources that are networked on Web-based systems. Most library Web sites have to be upgraded to serve as useful portals. A rigorous information literacy campaign will also be needed.

Six of the eight programs used library automation package in the instruction of courses. Four of them accessed the automation package of Horizon while one used Unicorn. At SQU they were in the process of converting from the locally developed system of Afaq to Unicorn. UQU Women also used the digital library system of Dspace. Arabian Advanced System has been championing the use and customization of Horizon as a bilingual integrated system. So many library systems in this region are using this system, facilitating networking and resource-sharing prospect. Another encouraging development is that OCLC has developed a sizable Arabic language database that may also serve as a source of authority control and standardization for managing bibliographic systems, utilities, and library catalogs.

One important instructional resource is the use of bibliographic databases for search and retrieval and research. Among the databases that were accessible to these programs, all the eight accessed LISA and ERIC. Six of them had access to Academic Search, ABI Inform, and Dissertation Abstracts Online. Five had access to Library Literature. Four accessed Emerald Full-text, Ulrich Plus, and General Science Index. Three of them were found to be accessing Encyclopedia Britannica and Readers Guide. Two of them reported access to Web Dewey, Classification Web, and ISI Web of Knowledge. KU's program reported access to BIP, PsychInfo, and LISA.

15.9 Evaluation Practices

The primary purpose of this chapter is to review evaluation practices prevalent in the LIS programs in the six GCC member nations. Two modes of evaluation were identified for this study—self-study and evaluation by external reviewers. In the following section, we have analyzed evaluation strategies and practices of these programs in relation to self-study and external review.

Table 15.3 Self-study

Institution	KAU	QU	SQU	UQU women	KU	PAAET	KSU	UQU men
Year of self-study	2005	?	2003–2004	None	2001–2002	2000	None	2004
Strategic plan		X	X		X	X		X
Students		X	X		X	X		X
Curriculum	X	X	X		X	X		X
Student evaluations		X			X	X		X
Faculty		X	X		X	X		X
Research output		X			X	X		X
Academic management		X	X			X		X
Computing facilities	X	X	X		X	X		X
Library resources		X			X	X		X
Market needs		X	X			X		X
Survey of graduates		X	X			X		

15.9.1 Self-Study

We used the following eleven variables as criteria for assessing the coverage of self-study and review by external consultants:

1. Strategic plan
2. Student enrollment
3. Curriculum
4. Student assessment
5. Faculty
6. Research productivity
7. Academic management
8. Computing facilities
9. Library resources
10. Market needs
11. Survey of graduates

Table 15.3 shows results about self-study in the eight programs. UQU Women and KSU had not conducted self-study. Five programs gave the date of their last self-study exercise. Accordingly, PAEET program had conducted self-study in 2000, KU in 2001, 2006/2007, and 2012, SQU in 2003–2004, UQU Men in 2004, and KAU in 2005.

QU and PAAET programs claimed having covered all listed aspects. UQU Men had examined all the variables except conducting survey of graduates. KU did not cover surveys of graduates and market needs and managerial aspects, but it has conducted an extensive survey of the market needs and graduate perceptions for the self-study exercises of 2006 and 2012. KAU had only reviewed curriculum and computing facilities. Similarly, SQU program did not cover student evaluations and library resources in its self-study exercise. It appears that most of the programs conducted self-study in a thorough and comprehensive manner.

Most of the LIS programs were conscious about the value of self-study and they conducted in a meaningful way. Such reports are antecedent to the visit of external consultants, as there is no accreditation mechanism in place in region. Later we have analyzed how did these programs benefit from this exercise as they introduced a number of changes in their policies and strategies, pursuant to self-study.

15.9.2   External Review

We used the same 11 variables for examining the conduct of evaluation through external review. It was pertinent to know who conducted the review and when was it conducted. Table 15.4 shows results of external review in the eight programs, using the same criteria. It was found that both the Men and Women wings of UQU had not conducted external evaluation. Five of them had used an external expert for review whereas an appointee of the Ministry of Education examined the KSU program. At KSU and QU, external assessors focused only on one of the variables of curriculum in 2006 and 2007 respectively. PAEET and SQU programs were evaluated in the respective years of 2000 and 2007 for all the 11 variables. KAU's review, conducted in 2007, examined the aspects except student perceptions. In 2002, 2007, and 2012, KU program was examined for all variables except the two variables of academic management and student perceptions.

External consultants normally present essence of their judgments and recommendations in their final meetings with the top academic management. This may carry certain leverage for the academic programs when the academic managers of these programs pursue changes at a later stage. Also it carries additional dividend if the report is summarized and presented in different forums of academic hierarchy of an institution.

15.10   Outcome of Evaluation

No matter, how evaluation is conducted, it is important that the findings of this exercise are translated into useful and meaningful changes. This is the basic objective of evaluation exercise. We also described if the review exercise in the academic programs of the GCC had resulted in any changes during the 5 years following evaluation. Outcome was identified through definition of new strategic plan, changes in admission or graduation policies, changes in the provision of resources and facilities, changes in instructional approaches, and changes in curriculum. The responses are displayed in Table 15.5.

Table 15.4 External review

Institution	KAU	QU	SQU	UQU women	KU	PAAET	KSU	UQU men
Year of self-study	2007	?	2007	None	2002–2003	2000	2006	None
Who conducted?	Outside expert	Outside expert	Outside expert		Two outside experts	Outside experts	Appointee of the Ministry	None
Strategic plan	X		X		X	X		
Students	X		X		X	X		
Curriculum	X	X	X		X	X	X	
Student evaluation	X		X		X	X		
Faculty	X		X		X	X		
Research output	X		X		X	X		
Academic management	X		X			X		
Computer facilities	X		X		X	X		
Library resources	X		X		X	X		
Market needs	X		X			X		
Survey of graduates			X			X		

Table 15.5 Outcome of review exercises

Institution	New strategic plan	Changes in admission/graduation policies	Changes in resources and facilities	Changes in instructional approaches, etc.	Changes in curriculum			
					Whole curriculum	Course adjustments	Changes in required courses	Changes in elective courses
KAU	None	None	None	None	None	None	None	None
QU	X	X	None	None	X	X	X	X
SQU	None	None	None	None	None	X	X	None
UQU women	None	X	None	None	X	None	None	None
KU	X	X	None	None	X	X New courses	None	None
PAAET	None	None	None	None	None	None	None	None
KSU	None	X	None	None	X	None	None	None
UQU men	X	X	None	None	X	None	None	None

PAEET and KAU reported no changes in the five areas in 5-year span. KU had a new strategic plan, it had introduced changes in graduation requirements, and it revamped its curriculum and added new courses. QU had a new strategic plan and there had been changes in the admission and graduation policies. SQU made adjustments in coursework and some changes in required courses. In both the Men and Women wings of the UQU, changes in policies for student admission were made whereby they decided to admit only those students who had majored in science subjects in their high school. Both the wings had revamped their curriculum and it was oriented to information science. The Women wing also reported that it had a new strategic plan. At KSU, they had introduced changes in their admission/graduation policies. Also, they made changes in curriculum.

From this description, we have been able to ascertain the value of quality assurance and evaluation programs for the academic programs of this region. However, transferability of skill and employability of graduates was only a practical proposition in Saudi Arabia where the public sector is the primary employer and this sector did not discriminate on the basis of the academic programs a student graduated from. Other five states in the GCC are small entities and each of them has a unique administrative setup, defying any mutual transfer of skilled workforce. Thus recognition of degrees across the region does not have serious implications. However, if an individual with educational qualification of another country happens to be in another sister nation, there is recognition of the degree. Yet, at present, there is no large-scale transferability of academic credentials.

15.11 Accreditation: Rationale and Significance

Rehman (2012) studied accreditation perspectives of academic managers of LIS programs in the region and proposed a workable strategy for implementing accreditation system in the region. The author's ideas on the subject were published recently and these ideas have been used in our discussions on accreditation.

Accreditation has played a critical role in evaluating LIS programs on a periodic basis, bringing a sense of uniformity and standardization in LIS education, and cultivating professional knowledge, skills, capabilities, and values among professionals joining the workforce. ALA accreditation standards (1992) defined accreditation as follows:

Accreditation assures the educational community, the general public, and other agencies or organizations that an institution or program (a) has clearly defined and educationally appropriate objectives, (b) maintains conditions under which their achievement can reasonably be expected, (c) is in fact accomplishing them substantially, and (d) can be expected to continue to do so. Accreditation serves as a mechanism for quality assessment and quality enhancement with quality defined as the effective utilization of resources to achieve appropriate educational objectives [1].

Majid, Chaudhry, Foo, and Logan (2003) had a more straightforward definition of the term as a process which assures that educational institutions and their programs meet appropriate standards of quality and integrity. It is a collegial process based on self-evaluation and peer assessment for the improvement of academic quality and public accountability. ALA has been a pioneering agency in accrediting LIS programs in North America. Its Committee on Accreditation has been responsible for assessing Master degree programs offered in the North American schools every seventh year. ALA has been responsible for issuing standards, guidelines, and other instruments used in the accreditation process. ALA-accredited degree has thus become a norm for employment of LIS professionals in North America.

Other national and international agencies have also made efforts in proposing accreditation standards and guidelines. Most noteworthy among them are ALIA, CILIP, and IFLA. ALIA and CILIP: They have been responsible for assessing LIS education programs in Australia and the UK. ALIA's process is labeled as course recognition. One distinctive feature of these accrediting agencies is that they have covered undergraduate and graduate degrees, consistent with the professional practices prevalent in the two countries. IFLA, being an international forum, proposed its own accreditation guidelines. The Web sites containing accreditation documentation of these four agencies are as follows:

ALA Standards for accreditation of Master's programs in library and information studies[1]

IFLA Guidelines for professional library/information educational programs–2000[2]

ALIA Education policy statement nº 1[3]

ALIA The library and information sector: core knowledge, skills and attributes[4]

CILIP Accreditation instrument: Procedures for the accreditation of, courses[5]

Khoo, Majid, and Chaudhry (2003) examined the documentation of these agencies and made useful comparisons. The most important finding is that these accreditation standards focus on the following areas in their assessment:

• The context of the program, institutional support, and relationship with the parent institution
• Mission, goals and objectives
• Curriculum
• Faculty and staff
• Students
• Administration and financial support
• Instructional resources and facilities
• Regular review of the program, the curriculum, and the employment market

---

[1] http://www.ala.org/Content/NavigationMenu/Our_Association/Offices/Accreditation1/pub/standards.htm.

[2] http://www.ifla.org.sg/VII/s23/bulletin/guidelines.htm.

[3] http://www.alia.org.au/policies/education/entry-level.courses.html.

[4] http://www.alia.org.au/policies/core.knowledge.html.

[5] http://www.cilip.org.uk/NR/rdonlyres/AB7FB628-3922-4681-85AA-3E75593A0389/0/ACCREDITATIONWEB.pdf.

- Documentation

These associations have also tried to define the core competencies of professionals that the LIS programs need to focus in their curricula and educational thrust. ALA had initiated definition of core competencies in 1999. The draft was presented to a number of committees and conferences for review. In 2005, the exercise resulted in a document that outlined core competencies. McKinney used these statements for examining the curricula of accredited, accreditation-candidates, and pre-candidates. The core competencies, yet to be formally adopted by ALA, were defined as follows:

1. Professional ethics
2. Resource building
3. Knowledge organization
4. Technological knowledge
5. Knowledge dissemination: service
6. Knowledge accumulation: education and life-long learning
7. Knowledge inquiry: research
8. Institution management

After examining curricula and syllabi of 58 LIS program, McKinney found that knowledge organization, professional ethics, knowledge dissemination, technological knowledge, research, and management competencies were covered in the required coursework of 53, 45, 41, 37, 37, and 36 programs respectively. All the core competencies were however covered in the required and elective coursework. Results of this study indicated that the accredited schools in North America had an adequate coverage and treatment of the core competencies. That validates the relevance and values of these competencies. Core skills and competencies specified in CILIP and ALIA are quite detailed. IFLA guidelines list eleven areas, which are as follows:

1. The Information Environment, Information Policy and Ethics, the History of the Field
2. Information Generation, Communication, and Use
3. Assessing Information Needs and Designing Responsive Services
4. The Information Transfer Process
5. Organization, Retrieval, Preservation, and Conservation of Information
6. Research, Analysis, and Interpretation of Information
7. Applications of Information and Communication Technologies to Library and Information Products and Services
8. Information Resource Management and Knowledge Management
9. Management of Information Agencies
10. Quantitative and Qualitative Evaluation of Outcomes of Information and Library Use

The LIS schools that are candidates for accreditation have to go through a self-study exercise. These also need to develop adequate documentation for accreditation team. The process requires that the schools engage in an intensive exercise, which entails dialog and collaboration with many stakeholders such as faculty members, academic management of the parent institution, professional bodies, students and alumni, accreditation body, and other schools and programs in the region. Majid et al. (2003) listed following problems Southeastern nations may face, using the umbrella of CONSAL.

- Non-availability of funds
- Limited understanding and appreciation
- Lack of experts in developing and implementing accreditation
- Procedural difficulties
- Resistance form the LIS programs
- Fear of being exposed
- Government rules and regulations

Accreditation adds value to the efforts of LIS programs in keeping themselves current and relevant. Changes in the LIS profession have been swift and these need to be reflected in academic policies and curricula. Accreditation is a source of authentication for the accredited programs.

15.12 Perceptions About Accreditation

Consistent with the universal recognition of the need of accreditation, we also examined what were the perceptions of academic managers of LIS program in the GCC about this phenomenon and what were the possibilities for introducing a system of accreditation in this region. All the academic managers were affirmative about the value of accreditation. They gave their input about the possible accrediting agency that could manage the process. Seven of them considered a regional professional body such as SLA/AGC or a new body in the region that may assume this responsibility. Three of them considered the national professional association, as the right forum while another three thought the regional consortium of universities should conduct evaluation. Only one of them marked the choice of an international agency.

With regard to quality assurance aspect of uniformity and transferability of workforce, all of them were of the view that in this region, students should be uniformly accepted for further admissions, academic transfers and employment.

Saudi Arabia has a national agency named Saudi National Commission for Academic Accreditation and Assessment. This agency caters for all the disciplines in which degrees are offered in Saudi Arabian universities. The agency may ensure that suitable arrangements of accreditation are made for different disciplines. They may not have a pool of experts who can serve all the areas (Table 15.6).

Table 15.6 Perceptions about accreditation

Institution	Uniformity	Need for changes in assessment procedures	Accreditation	
KAU	Yes	Yes, using the Saudi National Commission for Academic Accreditation and Assessment	Yes	2, 3
QU	Yes	Yes, it should be done by the department and not by the college	Yes	2, 3, 4, 5
SQU	No	No	Yes	2
UQU women	Yes through regional accreditation agency	Yes, invest in the process and reward those who perform it	Yes	2, 5
KU	Yes	No	Yes	2
PAAET	Yes through GCC certification process	Yes, it should be conducted every 5 years	Yes	1, 2, 3, 4
KSU	No	Yes	Yes	1
UQU men	Yes	Yes, Invest resources and rewards those who work on it	Yes	2, 5

15.13   Accreditation Proposal for the Region

Since ALA or other national or international professional bodies do not accredit programs in other countries, it is not an option. An alternate suggestion is using a regional professional body such as SLA's Arabian Gulf chapter. SLA does not engage in accreditation activities. Also, the Arabian Gulf Chapter is a loosely structured forum that does not have any headquarters or permanent staff. At present, this is the only regional forum that has been holding a conference annually. Yet it cannot be given a task for which it is not prepared. Likewise Ministries of Education cannot have the capability to conduct accreditation reviews. Since many nations do not have national professional societies or associations, this may also not be practical. The largest country in the group is Saudi Arabia. It did not allow creation of professional forums. Recently the country has relaxed its society laws, but the society created is not mature or strong. It would take a long time for any of these associations to mature enough for such an undertaking.

One strong view is that the regional consortium of universities may be entrusted with this task. Within the framework of GCC, there is a permanent body of GCC Universities and the executive heads of these universities meet regularly. Within that structure there is a permanent forum of deans of libraries in the region. If the GCC Universities forum is approached to constitute an organ of the heads of academic departments of LIS for accreditation, this may be given this responsibility. It may function both in statutory and professional capacities. This forum may have the responsibility of policy formulation, development of guidelines and instruments, financial management, constitution of accreditation teams, etc. An initiative on the part of academic departments is needed if any such proposal is to be tabled before the forum of the chief executives of the Gulf Universities.

15.13.1 Accreditation Team

The forum proposed in the preceding section might be responsible for the constitution of accreditation team. The appointment of a member could be made for a certain period. It is proposed that the team should have eminent educationists from the region and about as many members might be picked from the international market. This might be a sensitive issue for the academic departments, but the practices of ALA, ALIA, and CILIP might provide guidelines.

One possibility could be to explore with IFLA if it could have a stake in the process.

15.13.2 Standards and Guidelines

Once the accrediting agency is in place, it should engage senior academics to draft standards, guidelines, instruments, and processes. There should be a representation of all the schools for this task force. Detailed documentation is available from the afore-cited Web sites of four professional associations. Appropriate adjustments and customization would be desirable.

There appears to be a common core of elements that are to be evaluated in the accreditation process. It has been found that all these elements have been covered in the self-study or external review exercises of the nine LIS departments surveyed.

15.13.3 Obstacles in the Accreditation Process

A number of challenges and obstacles are expected in the process of instituting an accreditation program in the region. Establishing an accreditation process will not be an easy process. In this region, LIS education is primarily conducted at the undergraduate level. There are only two structured Master degree program in Kuwait and UAE. ALA accredits only Master degree programs. CILIP and ALIA programs are however oriented to both the levels of education. Another issue is that two undergraduate majors and one undergraduate minor have little to do with librarianship.

15.14 Points to Ponder

One major issue in the region is that the salary structures and position classifications of LIS professionals, enforced by the civil service authorities in these countries, provide that the LIS professionals should have an undergraduate degree. During the last few decades, the professional education has shifted to graduate education worldover. However, in this region, most programs developed during the last three decades have a different setup of intake and graduation of LIS professionals. As a result, we find that the professionals in this region are deeply concerned about their status and their overall image in the society. A great deal of that might be attributed to the education and preparation of these professionals. Educators in this region need to critically examine this situation and find out what strategies might be appropriate to bring the education and preparation of professional at par with the international trends. Undergraduate education might be more appropriate for the purposes of creating information literacy and having workforce for vocational and technical jobs.

Another point of apparent concern is that most schools in this region have a very high student faculty ratio. The intakes and enrollments in these programs are high, but the number of faculty members is relatively modest or weak. In order to create conducive learning culture and having a meaningful engagement between the teacher and the taught, this ratio should be brought down to the global norms. The ratios of 1:30 or above pose serious problems and need to be brought down. It has also been noted that in Saudi Arabia and many other countries, employment of graduates has been a serious problem. The academics need to revise their admission and intake policies and strike a balance between supply and demand.

It has been found that monographic and periodical subscriptions in many of these programs are on the weak side. Having very few periodical subscriptions in three schools indicates a serious paucity of serial collection. It seems that these schools have little to encourage their faculty members for research as their collections are weak and student teacher ratios are high. Only one program with a graduate degree had 80 periodical subscriptions in English language. All others programs had less than 25 subscriptions, which might be an indicator of low value attached with research and scholarship. This study did not take into consideration research productivity of the faculty members, but an earlier study had indicated that the research productivity was low among the faculty (Al-Ansari, Rehman, & Yousef, 2001).

The situation of resources also deserves focus. Three schools were found to be having classrooms that were not electronically prepared. Some schools were subscribing to two databases. The fact that only two schools subscribed to Web Dewey and Classification Web also indicates that these might still be using the print version. In two programs, a small number of computing workstations were found whereas the number of students exceeded 500. All these factors indicate that the policy makers need to pay serious attention to the availability of resources and facilities that are crucial for making learning effective.

Policies and practices of evaluation through self-study or external reviewer largely vary among these programs. Using results of evaluation exercises will improve their situation. It was found that almost all the programs had used either of the two strategies for evaluation during the last 5–7 years. One school had a policy of a 5-year evaluation cycle. Five programs reported that they had used the results of this exercise for overall curriculum revamping and other adjustments. However, none of them had used these evaluation exercises for improving their computing facilities and instructional resources. The result of this study may provide a better insight to the academic policy makers to attend to the areas in which they can use the results of these evaluation exercises. It is worth mentioning that two programs have switched over to the information domain from the traditional LIS orientation in their academic programs. It has yet to be seen how their graduates make a mark in employment market in the near future.

Five program managers favored to have uniform policies for student intake, acceptance, credit transfer, and employability among these schools. The schools that did not favor had already shifted to information science and systems and did not feel comfortable that their curriculum, graduation requirements, and employability of graduates would permit a uniform treatment.

All the programs favored that an accreditation system should be in place. However, there was little agreement who should be the accrediting them. However, the largest number marked the option of a regional body such as SLA/Arabian Gulf chapter. Keeping in view that the largest number of graduates are employed in public sector schools, this recommendation is quite viable. It may not be as easy to propose an acceptable system of accreditation; it is only through continued interaction and engagement of stakeholders that these programs may approach a common ground of understanding.

References

Abell, A., (1998). Editorial. *Journal of Librarianship and Information Science, 30*(4), 211–214.

Al-Ansari, H., Rehman, S., & Yousef, N. (2001). Faculty in the Library Schools of the Gulf Cooperation Council Member Nations: An evaluation. *Libri, 51*, 173–181.

Aman, M., & Mika, J. (2004). Developing a bachelor's degree program in library and information science in United Arab Emirates. *Journal of Education for Library and Information Science, 45*(3), 255–262.

American Library Association. (1992). *Standards for accreditation of Master's programs in library and information studies*. Chicago, IL: The Association Retrieved June 20, 2013, from http://www.ala.org/Content/NavigationMenu/Our_Association/Offices/Accreditation1/pub/standards.htm.

Blankson-Hemans, L., & Hibberd, H. (2004). An assessment of LIS curriculum and the field of practice in the commercial sector. *New Library World, 105*(7/8), 269–280.

Edomi, E., & Ogbomo, M. (2001). Career aspirations of master's degree students at the African Regional Center for Information Science (ARCIS), University of Ibadan, Nigeria. *Information Development, 17*(4), 262–267.

Genoni, P., Exon, M., & Farrelly, K. (2000). Graduate employment for qualifying library and records management courses at Curtin University of Technology. *The Australian Library Journal, 50*, 247–258.

Genoni, P., & Smith, K. (2005). Results of a longitudinal study of employment outcomes for Australian LIS graduates. Paper presented to *71st IFLA General conference and council* (pp. 14–18), held at Oslo, August, 2005.

Holma, B., & Pakalna, D. (2007). Aspects of the quality management of the LIS education. Paper presented to *INFORUM 2007: 13th Conference on professional information resources* (pp. 22–24), Prague, May. 2007.

Jefferson, R., & Contreras, S. (2005). Ethical perspectives of library and information science graduate students in the United States. *New Library World, 106*, 58–66.

KALIPER. (2000). *Educating library and information science professionals for a new century: KALIPER report*. Reston, VA: ALISE. Retrieved, from http://www.alise.org/publications/kaliper.pdf.

Khoo, C., Majid, S., & Chaudhry, A. S. (2003). Developing an accreditation system for LIS professional education programmes in Southeast Asia: Issues and perspectives. *Malaysian Journal of Library & Information Science, 8*, 131–149.

Loughridge, B., & Speight, B. (1996). Career development: follow-up studies of Sheffield MA graduates 1985/86 to 1992/93. *Journal of Education for Library and Information Science, 28*(2), 105–117.

Majid, S., Chaudhry, A. S., Foo, C., & Logan, E. (2003). "Accreditation of library and Information studies programmes in Southeast Asia: A proposed model. *Singapore Journal of Library & Information Management, 32*, 58–69.

McKinney, R. D. (2006). *Draft proposed ALA core competencies compared to ALA-accredited, candidate, and precandidate program curricula: a preliminary analysis issued February 3, 2006.* Revised and reissued February 10, 2006. Retrieved June 12, 2008, from http://www.ala.org/ala/accreditationb/Core_Competencies_Comparison.pdf.

Mohai, L. O. A. (1999). Employers' perceptions of the graduates and curriculum of a library school in Botswana. *Libri, 49*, 1–6.

Rehman, S. (2008). Quality assurance and LIS education in Gulf Cooperation Council (GCC) countries. *New Library World, 109*(7/8), 366–382.

Rehman, S. (2012). Accreditation of library and information science programmes in the Gulf Cooperation Council nations. *Journal of Librarianship and Information Science, 44*(1), 65–72. doi:10.1177/0961000611427723.

Saladyanant, T. (2006). Quality assurance of information science program: Chiang Mai University. In C. Khoo, D. Singh, & A. S. Chaudhry (Eds.), *Proceedings of the Asia-Pacific conference on library & information education & practice 2006 (A-LIEP 2006).* Singapore: School of Communication & Information, Nanyang Technological University.

Sarkhel, J. K. (2006). Quality assurance and accreditation of LIS education in Indian universities: Issues and perspectives. In C. Khoo, D. Singh, & A. S. Chaudhry (Eds.), *Proceedings of the Asia-Pacific Conference on Library & Information Education & Practice 2006 (A-LIEP 2006), Singapore, April 3–6, 2006* (pp. 427–431). Singapore: School of Communication & Information, Nanyang Technological University.

Tammaro, A. M. (2005). *Report on quality assurance models in LIS programs.* IFLA Education and Training Division. Retrieved June 12, 2013, from http://www.ifla.org/VII/s23/index.htm.

Tammaro, A. M. (2007). Performance indicators in library and information science (LIS) education: towards cross-border quality assurance in Europe. Retrieved June 12, 2013, from http://www.cbpq.qc.ca/congres/congres2007/Actes/Tammaro.pdf.

TFPL. (1999). *Skills for knowledge management; a report by TFPL Ltd.* London: TFPL.

Wozniczka-Paruzel, B. (2003). Experiences of library and information science (LIS) studies accreditation in the context of quality assurance systems in Poland. *Education for Information, 21*(1), 49–57.

Yen, D. C., Chen, H-G., Lee, S., & Koh, S. (2003). Differences in perceptions of IS knowledge and skills between academia and industry: Findings from Taiwan. *International Journal of Information Management, 23*(6), 507–522.

Zainab, A. N., Edzan, N. N., & Siti Suzana Abdul Rahman. (2004). Tracing graduates to ascertain curriculum relevance. *Malaysian Journal of Library and Information Science, 9*(1), 27–37.

Index

M. Miwa and S. Miyahara (eds.), *Quality Assurance in LIS Education:*
An International and Comparative Study, DOI 10.1007/978-1-4614-6495-2,
© Springer Science+Business Media New York 2015